The Philosophy of Luck

⚛ METAPHILOSOPHY

METAPHILOSOPHY SERIES IN PHILOSOPHY

Series Editors: Armen T. Marsoobian and Eric Cavallero

The Philosophy of Interpretation, edited by Joseph Margolis and Tom Rockmore (2000)

Global Justice, edited by Thomas W. Pogge (2001)

Cyberphilosophy: The Intersection of Computing and Philosophy, edited by James H. Moor and Terrell Ward Bynum (2002)

Moral and Epistemic Virtues, edited by Michael Brady and Duncan Pritchard (2003)

The Range of Pragmatism and the Limits of Philosophy, edited by Richard Shusterman (2004)

The Philosophical Challenge of September 11, edited by Tom Rockmore, Joseph Margolis, and Armen T. Marsoobian (2005)

Global Institutions and Responsibilities: Achieving Global Justice, edited by Christian Barry and Thomas W. Pogge (2005)

Genocide's Aftermath: Responsibility and Repair, edited by Claudia Card and Armen T. Marsoobian (2007)

Stem Cell Research: The Ethical Issues, edited by Lori Gruen, Laura Gravel, and Peter Singer (2007)

Cognitive Disability and Its Challenge to Moral Philosophy, edited by Eva Feder Kittay and Licia Carlson (2010)

Virtue and Vice, Moral and Epistemic, edited by Heather Battaly (2010)

Global Democracy and Exclusion, edited by Ronald Tinnevelt and Helder De Schutter (2010)

Putting Information First: Luciano Floridi and the Philosophy of Information, edited by Patrick Allo (2011)

The Pursuit of Philosophy: Some Cambridge Perspectives, edited by Alexis Papazoglou (2012)

Philosophical Engineering: Toward a Philosophy of the Web, edited by Harry Halpin and Alexandre Monnin (2014)

The Philosophy of Luck, edited by Duncan Pritchard and Lee John Whittington (2015)

The Philosophy of Luck

Edited by

Duncan Pritchard and Lee John Whittington

WILEY Blackwell

This edition first published 2015

First published as *Metaphilosophy* volume 45, nos. 4–5 (October 2014), except for "Luck as Risk and the Lack of Control Account of Luck," first published in *Metaphilosophy* volume 46, no. 1 (January 2015).

Registered Office
John Wiley & Sons Ltd, The Atrium, Southern Gate, Chichester, West Sussex, PO19 8SQ, UK

Editorial Offices
350 Main Street, Malden, MA 02148-5020, USA
9600 Garsington Road, Oxford, OX4 2DQ, UK
The Atrium, Southern Gate, Chichester, West Sussex, PO19 8SQ, UK

For details of our global editorial offices, for customer services, and for information about how to apply for permission to reuse the copyright material in this book please see our website at www.wiley.com/wiley-blackwell.

Library of Congress Cataloging-in-Publication data is available for this book.

ISBN 9781119030577 (paperback)

A catalogue record for this book is available from the British Library.

Cover image: Buster Keaton in a scene from *Steamboat Bill Jr.*, 1928. Buster Keaton Productions/RGA

Set in 10/11pt Times by Aptara Inc., New Delhi, India
Printed and bound in Malaysia by Vivar Printing Sdn Bhd

1   2015

CONTENTS

NOTES ON CONTRIBUTORS

Fernando Broncano-Berrocal is a postdoctoral fellow at the Centre for Logic and Analytic Philosophy, KU Leuven. He recently completed his Ph.D. at the LOGOS Research Group in Analytic Philosophy at the University of Barcelona, where he wrote a thesis on how accounts of luck bear on accounts of epistemic luck and in turn on theories of knowledge such as safety-based accounts and virtue epistemology. He has published adapted parts of his thesis in *Erkenntnis* and *Philosophia*.

E. J. Coffman is an associate professor of philosophy at the University of Tennessee. He works in epistemology and philosophy of action, and has recently published papers in *Philosophical Issues, Australasian Journal of Philosophy, Philosophers' Imprint, Philosophical Studies, Synthese, Philosophical Explorations*, and *Journal of Philosophical Research*. He is working on a book (to be published by Palgrave Macmillan) on the nature of luck and its bearing on the nature and scope of knowledge and free, accountable action.

Steven D. Hales is a professor of philosophy at Bloomsburg University of Pennsylvania. He writes primarily in epistemology and metaphysics, and his books include *This Is Philosophy* (Wiley-Blackwell, 2013), *Relativism and the Foundations of Philosophy* (MIT, 2006), *Nietzsche's Perspectivism* (co-authored, Illinois University Press, 2000), and *A Companion to Relativism* (edited, Wiley-Blackwell, 2011).

Jennifer Adrienne Johnson is an associate professor of psychology at Bloomsburg University of Pennsylvania. She earned her degree in experimental psychology with a concentration in cognitive neuroscience from McGill University. She has focused her research on better understanding human attentional processing and has published her work in journals such as *Cerebral Cortex* and the *Journal of Cognitive Neuroscience*.

Neil Levy is an Australian Research Council Future Fellow, based at the Florey Institute of Neuroscience and Mental Health in Melbourne, Australia. He works mainly on questions at the intersection of the sciences of the mind and ethics, as well as free will. His most recent books are *Hard Luck* (Oxford, 2011) and *Consciousness and Moral Responsibility* (Oxford, 2014).

Rachel McKinnon is an assistant professor of philosophy at the College of Charleston. Her main fields of research are epistemology and philosophy of language. Her primary focus is on the relationship between knowledge and action. Specifically, much of her research focuses on the norms of assertion. She has published widely on this topic in journals such as *Philosophical Studies, American Philosophical Quarterly*, and *Pacific Philosophical Quarterly*, and she is currently preparing a book manuscript.

Alfred R. Mele is the William H. and Lucyle T. Werkmeister Professor of Philosophy at Florida State University and director of the Philosophy and Science of Self-Control project (2014–2017). He is the author often books, including *Free Will and Luck* (Oxford, 2006), and more than two hundred articles.

Joe Milburn is about to complete his Ph.D. in philosophy at the University of Pittsburgh. He specializes in epistemology.

Duncan Pritchard is a professor of philosophy at the University of Edinburgh and director of the Eidyn Research Centre. His main field of research is epistemology, and he has published widely in this area, including the monographs *Epistemic Luck* (Oxford, 2005), *The Nature and Value of Knowledge* (co-authored, Oxford, 2010), and *Epistemological Disjunctivism* (Oxford, 2012).

Nicholas Rescher is Distinguished University Professor of Philosophy at the University of Pittsburgh, where he has served for over fifty years. He has published more than a hundred books ranging over virtually every branch of philosophy. The recipient of eight honorary doctorates from universities on three continents, he has served as president of the American Philosophical Association, the American Catholic Philosophical Association, the American Metaphysical Society, the American Leibniz Society, and the C. S. Peirce Society.

Wayne D. Riggs is an associate professor of philosophy at the University of Oklahoma. He works almost exclusively in the area of epistemology, though his philosophical interests range more widely, especially into the philosophy of science, value theory, and ethics.

Sabine Roeser is a professor of philosophy at TU Delft. Her research focuses on risk, emotion, and ethical intuitionism. She has published the monograph *Moral Emotions and Intuitions* (Palgrave Macmillan, 2011) and has edited numerous books on risk ethics plus a forthcoming volume, *Emotion and Value* (Oxford, co-edited with Cain Todd). She has published articles on risk and emotion in *Risk Analysis*, the *Journal of Risk Research*, and *Energy Policy*.

Lee John Whittington is about to complete his Ph.D. in philosophy at the University of Edinburgh. His thesis attempts to provide an account of the metaphysics of luck and show how this account applies to moral and epistemic luck.

INTRODUCTORY NOTE

DUNCAN PRITCHARD AND LEE JOHN WHITTINGTON

Luck is clearly an important philosophical notion. It plays a fundamental role in key debates across a number of core areas of philosophy, including epistemology (epistemic luck), political philosophy (just deserts), metaphysics (causation), and ethics (moral luck). And yet until quite recently the notion of luck was not itself explored by philosophers but rather was for the most part treated as an undefined primitive. Recent years have, however, seen an upsurge of critical exploration of luck in the philosophical literature. The goal of this collection is to bring together a broad cross-section of the best contemporary work on this topic, written by both established and up-and-coming philosophers.

The collection grew out of an international workshop on the philosophy of luck which was hosted by the Eidyn Research Centre at the University of Edinburgh in June 2013, and which was supported by funds from both Eidyn and the Scots Philosophical Association. This workshop featured talks by Nathan Ballantyne, Steven Hales, Joe Milburn, Wayne Riggs, Sabine Roeser, and Lee John Whittington. Many of the speakers at this event are represented in this collection (Steven Hales's contribution is now co-written with his colleague Jennifer Adrienne Johnson), though the essays here are inevitably very different from the versions presented in Edinburgh.

To the papers arising out of the workshop, we've added some further commissioned pieces in order to make this collection a more comprehensive treatment of the topic. In particular, we've included a new piece by Nicholas Rescher—who is arguably the first philosopher in the contemporary literature to put forward a theory of luck—that offers a useful overview

The Philosophy of Luck, First Edition. Edited by Duncan Pritchard and Lee John Whittington.
Chapters © 2014 Metaphilosophy LLC and John Wiley & Sons Ltd, except for "Luck as Risk and the Lack of Control Account of Luck" © 2015 Metaphilosophy LLC and John Wiley & Sons Ltd. Book compilation © 2015 Metaphilosophy LLC and John Wiley & Sons Ltd.
Published 2015 by John Wiley & Sons Ltd.

of his position. The collection also includes commissioned essays by several philosophers who have written distinctive work on luck in recent years: Fernando Broncano-Berrocal, E. J. Coffman, Neil Levy, Alfred Mele, and Rachel McKinnon. Finally, it includes a new statement and defence of the modal account of luck by one of the editors, Duncan Pritchard. The result is a collection which we believe substantially furthers the philosophical debate about the nature of luck, and which we hope will inform this debate in the years to come.

Finally, we would like to take this opportunity to thank the journal *Metaphilosophy* for including this volume in the *Metaphilosophy* Series in Philosophy. In particular, we would like to thank the editor in chief, Armen Marsoobian, for his support in this endeavour, and also the managing editor, Otto Bohlmann, for all his sterling work helping us to put the collection together.

CHAPTER 1

LUCK AS RISK AND THE LACK OF CONTROL ACCOUNT OF LUCK

FERNANDO BRONCANO-BERROCAL

The notions of luck and risk are closely related. Many of the luckiest events we can imagine occur in situations where there is a large amount of risk involved. Being the only survivor in a plane crash or winning roulette after betting one's life savings on one spin are examples of very lucky events that occur in situations where there is a lot of risk involved. In this essay, I aim to investigate the conceptual connection between luck and risk. More specifically, I aim to explain the former in terms of the latter.

Let me start with a quick overview of the philosophical literature on luck so as to frame the discussion. Philosophers who have theorized about luck have characterized the notion using three types of conditions: (1) *chance conditions*, (2) *lack of control conditions*, and (3) *significance conditions.* The core idea of chance conditions is that lucky events are by chance. Lack of control conditions roughly say that an event is lucky for an agent only if the agent lacks control over it, whereas significance conditions say that an event, even if chancy or beyond the agent's control, cannot be regarded as lucky unless it is significant to the agent.

Each type of condition has several versions, depending on how the relevant notions of chance, control, and significance are cashed out. For example, depending on what is thought to determine the chance of an event occurring, the chance condition is or might be formulated in terms of: (1.1) *accidentality* (the idea is that whatever makes an event accidental is what

The Philosophy of Luck, First Edition. Edited by Duncan Pritchard and Lee John Whittington.
Chapters © 2014 Metaphilosophy LLC and John Wiley & Sons Ltd, except for "Luck as Risk and the Lack of Control Account of Luck" © 2015 Metaphilosophy LLC and John Wiley & Sons Ltd. Book compilation © 2015 Metaphilosophy LLC and John Wiley & Sons Ltd.
Published 2015 by John Wiley & Sons Ltd.

makes it chancy);[1] (1.2) *indeterminacy* (the idea here is that chancy events are events that were not determined to occur prior to their occurrence);[2] (1.3) *subjective probability* (under this interpretation, the chance component of luck is cashed out in terms of what is expected to occur by the lucky agent);[3] (1.4) *epistemic probability* (this view states that chancy events are events that are not likely to occur given available evidence);[4] *the notion of being in a position to know* (according to this view, luck is a matter of failing to be in a position to know that an event will occur);[5] *objective probability* (chancy events are events whose occurrence is objectively unlikely);[6] and (1.7) *modality* (chancy events are events that would fail to occur in close possible worlds).[7]

On the other hand, lack of control conditions are formulated in terms of: (2.1) failing to exploit the target lucky event for some purpose (see Riggs 2009); (2.2) not being free to do something that would help produce it and prevent it (see Coffman 2009); and (2.3) there not being a basic action that the lucky agent could perform which she knows would bring about the lucky event (and how it would do so) (see Levy 2011, chap. 2).

Finally, the significance condition is stated in terms of: (3.1) the lucky agent being able to ascribe significance to the lucky event if she were availed of the relevant facts (see Pritchard 2005, 132–33); (3.2) the lucky event having some objective evaluative status for the lucky agent (a sentient being) (see Coffman 2007, 388); and (3.3) the lucky agent having some interest and the lucky event having some objectively positive or negative effect on it (see Ballantyne 2012, 331).

As one can imagine, different definitions of luck result from combining these versions of the significance, chance, and lack of control conditions. We do not need to go into further detail here. It suffices to note that some commentators think that the three types of conditions are necessary (and jointly sufficient) for luck,[8] that others drop the lack of control condition from their definitions,[9] that some others drop the chance condition and

---

[1] This view is endorsed by Rescher (1995, 32). See Latus (2003, 465, n. 17), Pritchard (2005, 127, 130; 2014), and Riggs (2009, 205) for relevant discussion. See also Unger (1968) for a related discussion in epistemology.

[2] See Coffman (2007, 389), Levy (2011, 14), Pritchard (2005, 126–27, 131), and Steglich-Petersen (2010, 363) for relevant discussion.

[3] Rescher (1995, 35) and Latus (2003) seem to uphold this view. See Steglich-Petersen (2010, 366–67) for relevant discussion.

[4] See Steglich-Petersen (2010, 367–68) for relevant discussion.

[5] Steglich-Petersen (2010) defends this view, which turns luck into an epistemic notion.

[6] Baumann (2012) upholds this view. Rescher (1995, 211–12 and passim) seems to uphold it too, although at some other places he seems to defend (1.3).

[7] Different formulations of modal chance conditions may be found in Coffman (2007, 390), Levy (2011, 19), and Pritchard (2005, 128; 2015).

[8] This is the case of Coffman (2007), Latus (2003), and, apparently, Levy (2011); Levy does not explicitly endorse a lack of control condition but nevertheless builds it into his proposed chance condition.

[9] This is the case with Baumann (2012), Pritchard (2005), and Rescher (1995).

go for a pure lack of control account of luck,[10] and that there are alternative accounts that include neither chance nor lack of control conditions.[11] It is also worth noting that most commentators think that the significance condition is necessary for luck, although this point has recently been disputed.[12]

With this overview of the literature in hand, we can now frame the forthcoming discussion. The aim of this essay is mainly positive— namely, to give an account of luck in terms of the notion of risk. This means that I will not argue against the definitions of luck referred to in the preceding paragraph. In addition, I assume (with most commentators) that a properly stated version of the significance condition is necessary for luck. My sympathies are with (3.3) (this is an assumption for which I will give no argument). Finally, I will only discuss those versions of the chance condition that I consider relevant to explaining luck in terms of risk. In particular, I will not address (1.1), (1.2), (1.3), (1.4), and (1.5). However, (1.6) and (1.7) will be relevant. As regards the lack of control condition, I aim to advance my own account of the notion of control. For reasons of space, (2.1), (2.2), and (2.3) will be only tangential to the discussion.

Without further ado, let me indicate how the essay is structured. In section 1, I make a distinction between two senses of risk that serve to account for luck: the event-relative and the agent-relative senses of risk. In section 2, I present two ways of interpreting event-relative risk: in modal and in probabilistic terms. In section 3, I argue that, although both interpretations of event-relative risk seem correct qua accounts of risk, only the modal interpretation correctly accounts for luck. In section 4, I cash out the agent-relative sense of risk in terms of lack of control and I suggest that any significant event that is lucky for an agent is an event with respect to which the agent is at risk. In section 5, I categorize lucky and nonlucky cases by means of the event-relative/agent-relative distinction concerning risk. As a result, I argue that there are two different types of luck (or of fortune): one that does not involve event-relative risk and one that does involve it. I argue that both arise out of agent-relative risk (that is, lack of control over the relevant event). In section 6, I advance an account of the notion of control so as to explain in which sense an agent is at risk with respect to an event and how this bears on luck. In section 7, I summarize my view and offer a reply to several counterexamples to the lack of control account of luck.

---

[10] This is the case with Riggs (2007 and 2009).

[11] See McKnnon (2013) for an alternative account according to which "skill is what we call the expected value of an ability, and luck is any deviation, whether positive or negative, from this value" (McKinnon 2013, 510).

[12] Pritchard (2015) points out: "We shouldn't expect an account of the metaphysics of lucky events to be responsive to such subjective factors as whether an event is the kind of thing that people care about enough *to regard as* lucky. That's just not part of the load that a metaphysical account of luck should be expected to carry."

1. Two Senses of Risk

The idea that luck can be explained in terms of risk is already in the literature. For example, E. J. Coffman interprets Duncan Pritchard's modal chance condition for luck (the condition that an event E is lucky for S only if E occurs in the actual world but would not occur in nearly, if not all, close possible worlds where the initial conditions for E are the same as in the actual world) (see Pritchard 2005, 128) as being equivalent to the view that luck is a matter of risk, where risk is understood in terms of the notion of easy possibility and easy possibility in terms of closeness between worlds, so that E could easily occur at t' just in case it would occur at t' in possible worlds close to actuality as they are at t (Coffman 2007, 390). Pritchard (2015 and manuscript) ratifies this interpretation by explicitly accounting for luck in terms of modal risk.[13]

Although the possibility of explaining luck in terms of risk is already entertained by Pritchard and Coffman, the considerations they offer on the connection between both notions are limited to only one sense of the notion of risk: the risk that lucky *events* had of not occurring. To be clear, there is a sense of the notion of risk that can be used to account for the notion of luck that has to do with the possible occurrence or nonoccurrence of an event, as when we say that there is risk that a World War II grenade on the tip of a cone will fall or that a talk to be given by a speaker with flu will be canceled. Let us call this sense of risk the *event-relative sense of risk.*

My claim (and here comes what I take to be a novel point in the literature) is that there is another interesting application of the notion of risk to the luck debate. In addition to affirming or denying that events are at risk of occurring or not occurring, we also affirm or deny that *agents* are at risk with respect to the possible occurrence or nonoccurrence of events, as when we say that a child is at risk with respect to the possible explosion of the World War II grenade which he has just picked up from the ground and with which he is enthusiastically playing, or less dramatically, the risk at which the members of an audience are with respect to the possible cancellation of the talk they are attending. Let us call this sense of risk the *agent-relative sense of risk.* My proposal is that when it comes to account for the notion of luck as a form of risk we should take into consideration not only the risk

[13] More precisely, Pritchard argues that there is a close relationship between the notions of luck and risk by showing that judgments about luck and risk go hand in hand. His point is that in order to account for this correlation risk must be defined in modal terms using a measure of closeness between possible worlds fixed by intuitive similarity (according to Pritchard, the alternative would be a probabilistic measure of closeness, but it would not serve to capture the sense of luck involved in lotteries). That said, Pritchard acknowledges two differences between luck and risk: (1) risk concerns unwanted events, while luck concerns both wanted and unwanted events; (2) we can meaningfully talk of very low levels of risk but not of luck—for example, he thinks that if a lottery is rigged to ensure that one is guaranteed to be the winner, it is not a matter of luck that one wins, but the case might still be described as a case of risk. In this essay, I address (1) and (2) (see, respectively, sections 3 and 5).

that lucky events had of not occurring (as Pritchard and Coffman do) but also the risk at which agents are with respect to lucky events. The task I undertake next is to give adequate definitions of the two senses of risk just distinguished and to show how they can be used to account for luck.

2. Event-Relative Risk: Modal and Probabilistic Interpretations

How can we understand the term "risk" in sentences of the form "There is risk that event E will occur" or "There is risk that event E will not occur"? One proposal is already offered by Coffman and Pritchard: we should understand it in terms of *close possibility of occurrence*. Keeping the spirit of their proposal, we can define the event-relative sense of risk as follows:

> *Modal Risk (MR)*: E is at risk of occurring at t if and only if E would occur at t in a large enough proportion of close possible worlds.[14]

Why does MR include the expression "in a large enough proportion of close possible worlds" rather than, say, "in most close possible worlds" or "in nearly all, if not all, close possible worlds"? Because there does not seem to be a fixed proportion of close possible worlds in which an event would have to occur to be considered, in all contexts, at risk of occurring. While it is true that in most situations we would not regard an event as being at a significant risk of occurring if it were not to occur in most or at least many close possible worlds (think about ordinary events such as snowing, running out of power while writing a paper, or bumping into a glass door), this is not true of all instances of event-relative risk. Some events are regarded as being at a significant risk of occurring even though they would fail to occur in most close possible worlds. Consider Nicholas Rescher's example of someone surviving a round of Russian roulette with one bullet in the chamber of a revolver with a six-shot capacity.[15] The approximately 0.16 probability of being shot indicates that the person would survive in most close possible worlds. This does not, however, prevent us from claiming that she was at a significant risk of being shot and hence of dying. Examples of this sort indicate that an event can be at a significant risk of occurring even though its occurrence is not *easily* possible.[16]

---

[14] For risk of nonoccurrence: E is at risk of not occurring at t if and only if E would not occur at t in a large enough proportion of close possible worlds. See Broncano-Berrocal 2014 for relevant discussion on the application of MR to the epistemic case.

[15] Rescher (1995, 25). See also Levy (2011, 16–18) for relevant discussion. Levy uses this example to argue that "the degree of chanciness necessary for an event to count as lucky is sensitive to the significance of the event" (2011, 17), so there is no fixed proportion of close possible worlds in which the event would have to occur to be considered lucky in the actual world, since the significance of the event may vary from case to case.

[16] This goes against Coffman's proposal that risk is a matter of there being an easy possibility that an event will occur (or will not occur, depending on the case).

An alternative to MR, which defines the event-relative sense of risk in terms of close possibility of occurrence, is to understand event-relative risk in terms of *high probability of occurrence*:

Probabilistic Risk (PR): E is at risk of occurring at t if and only if there is high probability that E will occur at t.[17]

The kind of probabilities that are relevant in PR are *physical probabilities* or *chances*, that is, the kind of probabilities posited by scientific theories, which are determined neither by scientific evidence nor by degrees of belief but by features of the world. This is why most philosophers call them *objective* probabilities. For example, the risk of developing cancer by having a diet based on processed food is greater than the risk of developing cancer by having a diet based on vegetables because the probability of the former is higher than the probability of the latter. PR explains risk in these terms.[18]

What is the correct way of conceptualizing risk in general: MR or PR? In my opinion, both are correct. PR better fits the notion of risk that is used in scientific and technical contexts, where the risk that an event has of occurring is usually determined with scientific models that calculate, for example, the objective probability of occurrence of that type of event in a long sequence of trials.

By contrast, MR better fits the ordinary notion of risk. In everyday life, when it comes to assessing the risk that an event has of occurring, we resort to our cognitive capacity to handle subjunctive conditionals. The judgments delivered by this capacity are arguably less precise than those delivered by a scientific model. What might be regarded as a defect, however, is in fact a virtue, because, on the other side of the coin, it allows us to make lots of true risk ascriptions quickly and on the basis of insufficient evidence, something that has an adaptive value.[19]

Therefore, PR and MR seem to capture two complementary sides of the event-relative sense of the notion of risk. As we will see next, they serve to formulate two different types of objective chance conditions for luck that roughly correspond to (1.6) and (1.7) (see the introductory section

---

[17] For risk of nonoccurrence: E is at risk of not occurring at t if and only if there is high probability that E will not occur at t (or low probability that E will occur at t).

[18] The probabilities that PR takes as values must be nontrivial (that is, probabilities other than 1 or 0). The reason is that if an event has probability 1 of occurring we would not say that it is at a high risk of occurring; we would say rather that it is a certainty that it will occur.

[19] Of course, this is not to say that we must avoid doing probabilistic calculations when it comes to assess the risk of a given situation. Probabilistic calculations can certainly help us determine whether there is risk that an event will happen, and, more important, they can guide us when our counterfactual capacities deliver incorrect judgments, as often happens in well-known errors, such as the gambler's fallacy. After all, our cognitive capacities are reliable but not infallible.

of the essay) and, in this sense, they help characterize the phenomenon of luck as an instance of the more general phenomenon of risk. Nevertheless, although the notions of risk that underlie these chance conditions seem both correct, I will show that only chance conditions formulated in modal terms serve to account for luck.[20] In particular, I will show that objective probabilistic chance conditions cannot account for cases of *highly probable lucky events.*

3. Event-Relative Risk: Modal or Probabilistic?

Intuitively, winning a fair lottery is at risk of not occurring in the sense stated by MR because one would not win in most close possible worlds (which is a large enough proportion). Winning a fair lottery is also a paradigmatic case of luck, so MR can be intuitively used to motivate corresponding modal chance conditions for luck like the following:

> *Modal Chance (MC)*: E is lucky for S only if E occurs in the actual world but would not occur in a large enough proportion of close possible worlds where the relevant initial conditions for E are the same as in the actual world.[21]

The same applies to PR. Intuitively, winning a fair lottery is at risk of not occurring in the probabilistic sense because, prior to winning, there was high probability that one would lose. PR motivates corresponding probabilistic chance conditions for luck like the following:

> *Probabilistic Chance (PC)*: E is lucky for S at t only if, prior to the occurrence of E at t, there was low objective probability that E would occur at t.

One might prefer to use conditional probabilities to formulate the relevant chance condition:

> *Conditional Probabilistic Chance (CPC)*: E is lucky for S at t only if, prior to the occurrence of E at t, there was low objective probability conditional on C that E would occur at t.[22]

---

[20] Pritchard (manuscript) thinks that PR is not the correct way of conceiving of risk. To show that, he gives two examples that supposedly prove that two equivalent events with the same probability of occurrence might have different levels of risk. I do not share Pritchard's intuitions about the cases and I do think that PR is a legitimate and correct way of conceptualizing risk, but note that if it turns out that PR is an inadequate way of capturing the metaphysics of risk (as Pritchard argues), then that would count in favor of my argument that chance conditions modeled on PR do not serve to account for luck.

[21] MC is basically Pritchard's modal chance condition for luck (Pritchard 2005, 128) modified so that it includes Levy's phrasing "in a large enough proportion of close possible worlds" (Levy 2011, 17) for the reasons exposed above.

[22] Baumann (2012) endorses this kind of chance condition for luck.

C is whatever condition one uses to calculate the probability that E will occur. For example, the unconditional probability that Lionel Messi will score at the match is high, but given C (the fact that he is injured) the probability that he will score is low. Suppose that Messi ends up scoring by luck. CPC explains why: Messi was injured, and therefore, given his injury, it was not very probable that he would score.

Although objective probabilistic chance conditions explain most cases of luck, I will argue that there might be high (conditional or unconditional) objective probability that E will occur at t and yet E might occur by luck at t. In other words, I will show that there are highly probable lucky events and hence that PC and CPC are not necessary for luck. Consider the following hypothetical scenario:

> *Lazy Luke.* In a distant future, the galaxy is populated by billions and billions of people. The billions of corporations of the Galactic Empire are hiring computer technicians. Our hero, Luke, is an unemployed computer technician. He is extremely lazy and does not want to work at all. All he wants is to lie on the couch and play video games. The Galactic Empire's political system, however, forces unemployed people to apply for jobs constantly, so Luke reluctantly switches on his supercomputer and starts applying for billions of jobs. Luke, who is a clever guy after all, uploads a very bad CV to the system. In fact, he makes sure to upload the worst CV of the galaxy (he knows how to do that). Hiring decisions are made based on the number of candidates and the quality of their CVs, so by submitting a disastrous CV, Luke ensures that whenever there is another candidate, he will not be chosen. Furthermore, he knows that his name is on the I.D.L.E. list, that is, the list of Individuals Devoted to Leisure and Enjoyment, which contains the names of those who should never be hired because of their extreme laziness (all companies use I.D.L.E.). Competition for jobs is fierce, so for every single job there are millions of applications (something that Luke knows). He also knows that people normally inflate their CVs.
>
> Today everything seems to be alright: for each job offer, Luke has uploaded the worst CV, has checked that there are more applicants, that he is on I.D.L.E., and so on. Unbeknownst to him, however, there is a problem with the application sent to company No. 86792922, MicroCorp. Due to some unusual interference in the data stream, the contents of the CV Luke has sent to MicroCorp suddenly change in such a way that the human resources department receives a CV full of so many brilliant achievements that they decide to hire Luke instantly (by law, once a company hires a worker, the worker cannot be fired for a period of one year).

Intuitively, it is by bad luck that Luke gets the job (remember: he does *not* want to get a job). So it is by luck that he gets at least one job. PC says that an event (for example, getting a job) is lucky for an agent only if, prior to the occurrence of the event, there was low objective probability that the event would occur. In *Lazy Luke*, however, the probability of getting at least one job is very high. This is so because the probability of a disjunction of independent events (such as getting a job at different places) is $(Pr(E_1 \vee E_2 \vee \ldots E_n) = 1 - (Pr(\neg E_1) \cdot (Pr(\neg E_2) \cdot \ldots (Pr(\neg E_n)))$, which means that the bigger n is, the bigger will be the probability of the disjunction (that is, the probability that at least one of the disjuncts obtains). In *Lazy Luke*, n is huge, since the system allows Luke to apply for billions of jobs simultaneously. Thus, given the huge number of applications sent, the probability of getting at least one job is very high. It is by luck that Luke gets a job. Therefore, contrary to what PC requires, low probability of occurrence cannot be necessary for luck.

The result is no better in the case of conditional probabilities. Recall CPC: E is lucky for S at t only if, prior to E's occurrence at t, there was low objective probability conditional on C that E would occur at t, where C is some condition to be specified. In *Lazy Luke*, C might be the fact that, for each job, Luke makes sure (1) to upload the worst CV, (2) to check that there is a considerable number of applicants, (3) to check that he is on I.D.L.E., and so on. In the case of the job at MicroCorp, C could also include the fact that some interference in the data stream changes the contents of Luke's CV in such a way that the company receives an excellent CV. The probability that Luke gets a job at MicroCorp conditional on C (so understood) is certainly very low. But, once again, for a sufficient number of conditionalized disjuncts the probability that he gets at least one job is very high. Yet it is by luck that Luke gets a job at MicroCorp and, therefore, it is by luck that he gets at least one job. The condition stated by CPC is not necessary for luck.

By contrast, MC explains *Lazy Luke* in a rather simple, natural way: although the probability that Luke gets at least one job is high, it is by luck that Luke gets a job because in most close possible worlds in which he applies for the job offer at MicroCorp there is no data interference, and he does not finally get the job.

It is important to explain why, as in terms of MC, in close possible worlds Luke does not get a job in any other company (that is, why there is no close risk that he will be hired in any other place), given that he has applied for all job offers and hence given the high probability of getting at least one job. The reason is that for each job offer he *makes himself safe* from the eventuality of getting a job (he submits the worst CV, he checks that there are more candidates, that he is on I.D.L.E., and so on). His actual actions prevent possible worlds in which Luke would get a job from being close, because closeness is fixed by similarity to the actual world, and possible worlds in which he gets a job are dissimilar in that he does not perform such actions. Compare his situation with an inverse scenario: lottery players are not completely safe from the eventuality of winning (suppose the prize is death), despite the low probability of winning, because they are typically not in a position to make themselves safe from that eventuality, say, by rigging the lottery system. The point is that since typical lottery players do not *actually* rig the lottery, possible worlds in which they do not rig it are close, and in some of them they win, hence the risk of winning *despite its low probability of occurring*.[23] I conclude that the most adequate way to characterize the event-relative sense of risk that serves to conceptualize luck as risk is modal, not probabilistic.

4. Agent-Relative Risk as Lack of Control

Lazy Luke, a case of luck, involves not only the risk that Luke gets a job at MicroCorp (event-relative risk) but also the risk at which Luke himself is with respect to that eventuality. This is the agent-relative sense of risk, which, I argue, is an essential part of our understanding of luck. A definition of agent-relative risk should give an answer to the following question: What kind of relation must an agent have with an event in order for that event (its occurrence or nonoccurrence) to be safe for the agent? My proposal is that the relation is a relation of *control*: the occurrence (or nonoccurrence) of an event is safe for an agent just in case the event is under the control of the agent. Luke, for example, submits the worst CV in all job applications, he checks that there are more candidates, that he is on I.D.L.E., and so on. In this way, he has control over the eventuality of getting a job. The agent-relative sense of risk, that is, the risk at which an agent is with respect to a significant event can be accordingly defined as lack of control over the event.

One of my initial assumptions has been that the significance condition for luck is to be understood in terms of (3.3), the lucky agent having some interest and the lucky event having some objectively positive or negative effect on it.[24] The idea is that significant events are events that have a positive or negative impact on our subjective and/or objective interests. My suggestion is that in the same way as certain events are not lucky because they affect nobody's interests, one is not at risk with respect to certain events that are beyond one's control, precisely because they are not significant to one in the stipulated sense. For example, for most of us, the fall of a leaf in the middle of the Amazon jungle has no impact on our objective or subjective interests, as it affects neither our health nor our biological functioning, and has no effect on the objects of our desires or preferences. In the same way, we are at no risk with respect to several atomic nuclei joining in a very

---

[23] Here I am assuming what Williamson calls the "no close risk" conception of safety, which according to him "permits us to make ourselves safe from a disjunction of dangers by making ourselves safe from each disjunct separately, and to check that we are safe from the disjunction by checking that we are safe from each disjunct in turn" (Williamson 2009, 17). In *Lazy Luke*, the relevant disjunction is composed of all the job offers Luke applies for. He successfully makes himself safe from all disjuncts except one, the job at MicroCorp, hence his bad luck when he is hired.

[24] See Ballantyne (2012) for the original proposal.

distant point of the universe (something that we cannot control). We are certainly at risk, however, with respect to the same nuclear fusion if it triggers a nuclear explosion near us (something that we cannot control either). The reason is clear. In the latter case, the nuclear fusion affects our most important objective interest: being alive.

In view of these considerations, I propose the following definition of agent-relative risk:

> *Agential Risk (AR)*: S is at risk with respect to an event E if and only if (i) S has an interest N, (ii) if E were to occur, it would have some objectively positive or negative effect on N, and (iii) S lacks control over E.

My hypothesis is that any significant event that is lucky for an agent is also an event with respect to which the agent is at risk in the sense specified by AR. A qualification is in order, though. Lucky events negatively *or positively* affect one's interests. Compatibly, AR says that one is at risk with respect to whatever significant event is beyond one's control, which entails that one can be at risk with respect to events that decrease or increase one's well-being. In ordinary discourse, however, the term "risk" is most commonly used as synonymous with "danger" or "hazard," where a dangerous or a hazardous event is a significant event which has adverse or unwelcome consequences for one's interests and over which one lacks control. For example, the ordinary conception of risk as danger is the reason many would not say that lottery players are at risk of winning. But AR stretches the ordinary usage of the expression "at risk of" so that it does not necessarily mean "in danger of." Is this move justified? In which sense are we at risk with respect to uncontrolled significant events that end up increasing our well-being?

The short answer is that we are at risk with respect to those events because they affect us in ways that diverge from the path traced by our goal-directed controlled actions. When we have control over an event, we can count on it. However, events beyond our control, even those that carry positive effects, are events on which we cannot count, for example, in order to take further action. By way of illustration, inexperienced investors playing the stock market typically buy shares without knowing the relevant financial technicalities or the maneuvers that big investor groups perform to make money at their expense. Even if the prices of the shares rise and they become rich (which is something positive as far as their subjective interests are concerned), the rise of the prices is something on which they cannot count at the time of the investment because it is something beyond their control. Thus, it would be irrational for them to apply for a big loan from a bank to set up an expensive business on the assumption that the forthcoming profit in the stock market would be sufficient to repay it. AR allows for "good" risks, true, but risks after all, and in general we cannot rely on risks, even if unknowingly beneficial.

5. Four Combinations of Risks, Two Ways of Being Lucky (or Fortunate)

Let us put all the pieces together. On the one hand, an *event* can be at risk of occurring (or of not occurring). I have called this *event-relative risk*. On the other hand, an *agent* can be at risk with respect to an event. I have called this *agent-relative risk*. Event-relative risk has been understood in modal terms (MR), and agent-relative risk in terms of lack of control (AR). By combining these two senses of risk, we can come up with four different structures of cases. Suppose that an event *E* actually occurs. There are four possibilities:

A. *S* is at risk with respect to *E* and *E* was at risk of not occurring.
B. *S* is at risk with respect to *E* and *E* was not at risk of not occurring.
C. *S* is not at risk with respect to *E* and *E* was not at risk of not occurring.
D. *S* is not at risk with respect to *E* and *E* was at risk of not occurring.

A-cases constitute paradigmatic cases of luck. A good example of an A-case is winning a fair lottery. When one wins a fair lottery, prior to the lottery draw, the fact that one would win was at big risk of not occurring, that is, one would lose in close possible worlds (event-relative risk). In addition, if the lottery is fair, *one* is at risk of not winning because one has no control over the lottery process and its outcomes (agent-relative risk).[25]

B-cases also constitute cases of luck. A B-case would be, for example, a case in which one wins a lottery because, unbeknownst to one, the organizer has rigged the lottery in one's favor. In this case, there is no risk of losing the lottery, because the organizer diligently manipulates the lottery system so that one wins in the actual and in all close possible worlds (no event-relative risk). Still, *one* is at risk with respect to the outcome of the lottery because one has no control over the lottery process (agent-relative risk). For this reason, we intuitively say that one is lucky to win.

C-cases, by contrast, involve no risk whatsoever, so we should not expect the presence of luck in them. A C-case would be, for example, a case in which one rigs a lottery in one's favor. Winning in that case would not be by luck, because (1) the eventuality of losing is at no risk of happening (no event-relative risk) and (2) *one* is at no risk of losing, provided that one has control over the lottery process (no agent-relative risk).

D-cases are cases in which the relevant event is at risk of not occurring (or of occurring, depending on the case) and yet one is at no risk with respect to that event. Excellent examples of D-cases are decisions made on

---

[25] If one does not have control over the lottery process and its outcomes, one is not only at risk of losing but also at risk of winning (remember that AR can be applied to positive events as well). In general, we are attracted to lotteries because they allow us to expose ourselves intentionally to the whims of luck by engaging in a game whose relevant parameters are beyond our control. This lack of control gives us hope of winning even though we know that the probability of losing is extremely high.

a whim (*whimsical decisions*). Let me use one of Jennifer Lackey's examples (Lackey 2008). Suppose that one decides to go to Paris for the weekend on a whim. Since one has made the decision on a whim, the decision was at big risk of not being made (event-relative risk). However, *one* was at no risk of not deciding to go to Paris because, even though the decision was made on a whim and, therefore, one could easily have not made it, it was a self-consciously made decision after all, which means that one had control over the somehow precipitated deliberation process (no agent-relative risk). Interestingly, this means that, although one could easily have not made the decision, it is not by luck that one makes it.[26] In this sense, D-cases are not cases of luck.

Before using this taxonomy of cases to shed some light on the relationship between the notions of luck and risk, let me briefly evaluate Lackey's argument concerning *whimsical events* (significant events that result from whimsical decisions) to the conclusion that conjoining a modal chance condition (for example, MC) with a significance condition does not suffice to define the notion of luck. Keeping in mind that modal chance conditions roughly say that lucky events are modally fragile and that significance conditions say that they are significant, Lackey's argument is as follows:

(1) decisions made on a whim are modally fragile, that is, they would not occur in close possible worlds;
(2) given (1), if a significant event that results from a whimsical decision occurs in the actual world, the event would not occur in close possible worlds;
(3) significant whimsical events are not by luck;
(4) it follows from (2) and (3) that significant whimsical events are modally fragile but not by luck; and
(5) therefore, conjoining a modal chance condition with a significance condition does not suffice to define luck.

Although I agree with Lackey's conclusion (my account of luck is a version of the lack of control account of luck), I do not think that it follows from the premises. That is, one can accept (1)–(4) without accepting (5). Lackey's error in thinking that the conclusion follows is due to a misconception concerning the clause on initial conditions of MC and similar chance conditions. While it might be true that if one decides to go to Paris on a whim, one would not go to Paris in most close possible worlds, the only close possible worlds that are relevant to assess whether it is by luck that one goes to Paris in the actual world are worlds in which the relevant initial

---

[26] The same applies to *torn decisions*, which Mark Balaguer uses to defend a naturalistic libertarian account of free will. He defines torn decisions as the kind of decisions we sometimes make when we have reasons for two or more options and we feel torn as to which reason is the best, so we end up just choosing one of the options (Balaguer 2010, 71).

conditions for the occurrence of the event are the same as in the actual world. As a general rule, one's decision to φ is always among the relevant initial conditions for one's φ-ing. Therefore, close possible worlds in which the relevant initial conditions for going to Paris are the same as in the actual world are worlds in which one makes the decision to go to Paris. In all close possible worlds in which one decides to go to Paris, one goes to Paris. Consequently, MC and similar conditions do not hold, and, therefore, they correctly rule out one's going to Paris as a case of luck.

Let us return to how the taxonomy of cases above bears on the relationship between the notions of luck and risk. We will focus first on cases in which there is no luck involved, that is, on C-cases and D-cases. C-cases show that luck does not arise if there is neither event-relative nor agent-relative risk. D-cases further show that the absence of luck is compatible with there being event-relative risk. Together with C-cases, they also show that luck does not arise if there is no agent-relative risk, that is, the absence of agent-relative risk guarantees the absence of luck. According to AR, agent-relative risk is essentially a matter of lacking control over an event. Therefore, there is no agent-relative risk, and hence no luck if one does not lack control over the relevant event, in other words, if one *has* control over the event.

What about cases of luck (A-cases and B-cases)? In both there is agent-relative risk (that is, lack of control over the relevant events) and hence luck. The question is: Are the events of A-cases and B-cases lucky in the same way? From a theoretical perspective, they are not, as only A-cases involve event-relative risk. But the theory is also backed up by intuition: winning a fair lottery (an A-case) does not intuitively have the same quality of "luckiness" as winning a lottery that, unbeknownst to one, someone has firmly decided to rig in one's favor (a B-case). In other words, the different intuitions elicited by A-cases and B-cases point to the existence of two different senses of the notion of luck. The presence or absence of event-relative risk is what explains the difference.

The distinction has been noted already in the literature, although not in the same terms in which I introduce it here. In particular, Coffman points out that there is a difference between luck and fortune in the following way: "You can be fortunate with respect to an event whose occurrence was *extremely* likely, whereas an event is *lucky* for you only if there was a significant chance the event wouldn't occur" (Coffman 2007, 392).

What I have shown so far is a motivated way of arriving at the distinction underlying Coffman's quote by characterizing the phenomenon of luck as an instance of the more general phenomenon of risk. In a slogan, luck is just risk. More specifically, my view is that luck arises just in case an agent is at risk with respect to an event. But an agent's luck may come in two guises, depending on whether there is risk that the relevant event fails to occur. This is a real distinction, but, contrary to what Coffman thinks, I do not think that the terms "luck" and "fortune" capture it. For both terms can be

interchangeably used in ordinary discourse without risk of falsity or infe-
licitousness. For instance, we would indistinctively (and successfully) apply
the terms "luck" and "fortune" to both A-cases and B-cases, for example,
to both winning a fair and winning a rigged lottery.

Other commentators have also attempted to distinguish between luck
and fortune as if these terms captured a real distinction.[27] This is a mistake:
there is a real distinction, but it is not captured by the terms "luck" and
"fortune." Our ordinary usage of the terms overlooks the conceptual difference between A-cases and B-cases. In order to distinguish the two sides
of the concept to which the terms "luck" and "fortune" refer in ordinary
discourse, we can call *luck* or *fortune of type A* the kind of luck/fortune that
is present in A-cases and *luck* or *fortune of type B* the kind of luck/fortune
that is present in B-cases.

A qualification is in order. While there are clear-cut cases of A-luck (for
example, winning a fair lottery) and clear-cut cases of B-luck (for example, winning a lottery that, unbeknownst to one, has been rigged in one's
favor), there are some cases in which we do not know whether to ascribe
A-luck or B-luck. Contrary to what one might think, this is no objection
to the present account, as the ordinary concept of luck or fortune is inherently vague, and consequently we should not expect our analysis of luck
to remove all vagueness. On the contrary, it would be positive if it could
predict the existence of such cases. In other words, the distinction between
the two types (A-type and B-type) of luck or fortune need not be a sharp
distinction.

How can we know whether a case is borderline, that is, a case that is neither clearly A-lucky nor clearly B-lucky? In general, if a case is on the borderline between clear-cut A-luck and B-luck, we will be unable to appeal
to the modal fragility (A-luck) or modal robustness (B-luck) of the relevant event to explain its being by luck. Rather, the only thing to which we
will be able to appeal in order to explain its intuitive "luckiness" or "fortunateness" is the fact that the agent lacks control over the event. That is,
the somewhat vague limit between A-luck and B-luck (or A-fortune and
B-fortune) emerges when the lack of control intuition does by itself all the
explanatory work.

The concept to which the terms "luck" and "fortune" refer in ordinary discourse applies, therefore, to two different types of cases (A-cases
and B-cases), but also to cases on the borderline between them. Why is it
worth making the distinction? First, because it is a distinction we make at
the intuitive level when we judge the outcomes of fair and rigged lotteries

---

[27] For Pritchard (2005, 144, n. 15), fortunate events are events that count in one's favor
over which one has no control. In Pritchard 2015, he further specifies that fortune "tends to be
concerned with relatively long-standing and significant aspects of one's life, such as one's good
health or financial security" (2015). For Levy (2009, 495–97), fortunate events are nonchancy
events that have luck in their causal history (namely, in their proximate causes).

differently. If, unbeknownst to you, the organizer of a lottery has decided to make you win, your luck or fortune is not the same as when you win fairly. Second, it is worth making the distinction because it shows how pervasive the phenomenon of luck is.

To illustrate the latter point, consider Lackey's criticism concerning modal chance conditions such as MC (Lackey 2006; 2008). Lackey argues that cases like the following prove that they are not necessary for luck (adapted from Lackey 2006, 285):

> *Buried Treasure.* Sophie has the strong desire to bury a treasure at location *L*. There is no chance that she buries the treasure at any other place. She buries it. Vincent has the strong desire to place a rose bush in the ground of *L*. There is no chance that he places the plant anywhere else. He goes to place it. As he is digging, he discovers Sophie's buried treasure. Sophie and Vincent neither know each other nor know anything about the other's actions.[28]

Lackey argues that, while Vincent's discovery is intuitively by luck, it would occur in most close possible worlds, which proves that modal chance conditions are not necessary for luck. I agree with Lackey that *Buried Treasure* is a case of luck (or of fortune). No philosophical theorizing will be able to neutralize that intuition. But the absence of event-relative risk indicates that *Buried Treasure* is a case of type B (Vincent's actual discovery would still occur in most close possible worlds, that is, it is sufficiently modally robust, not chancy). However, modal *chance* conditions are not meant to be necessary for luck/fortune of type B. Rather, they help define luck of type A, that is, the kind of luck that involves event-relative risk, which we easily identify in fair lotteries and other gambling games. Luck, in this way, is a pervasive phenomenon that comes in two different guises.

The perceived luckiness or fortunateness in *Buried Treasure* is essentially explained by Vincent's lack of control over his discovery. Or would we say that Vincent's discovery is by luck if he had known that there was a treasure in the area and had used a metal detector? In the next section, I give a detailed account of the notion of control that aims to explain the sense in which we are at risk with respect to events beyond our control, and how this bears on luck. This allows me to specify in which way Vincent lacks control over his discovery.

6. An Account of the Notion of Control

In philosophy, the term "control" has been extensively used to account for a variety of concepts, such as action, property and ownership, freedom, privacy, personal autonomy, responsibility, and luck. Typically, it is assumed that we are all able to distinguish when things are under or beyond our

---

[28] Latus (2003, 468) and Rescher (1995, 35) give analogous examples.

control in such a way that control is regarded as an intuitive primitive notion to which one can resort to explain other concepts. Sometimes definitions or explications of control are given, but they normally aim to clarify its concrete role in the wider philosophical argument where the concept is used. Of course, there is nothing wrong with using the notion in some specific sense to serve some specific philosophical purpose, for example, to account for a special variety of luck—in fact, we cannot assume without argument that the special varieties of luck (moral, epistemic, distributive) can be defined with the same lack of control condition—but what we need here is a generic (yet detailed enough) account of the notion of control that may help us give an adequate account of the notion of luck as we ordinarily understand it.

Daniel Dennett, in one of his works devoted to the compatibility between determinism and free will, defines control as follows: "*A* controls *B* if and only if the relation between *A* and *B* is such that *A* can drive *B* into whichever of *B*'s, normal range of states *A* wants *B* to be in" (Dennett 1984, 52). Let me make three points about Dennett's definition. First, the definition indicates that, when we control something or someone, it is our intention, desire, goal, aim, target, plan, or purpose to achieve certain outcome concerning that thing or person.[29] Our interests, from the most basic ones (for example, self-preservation) to the most sophisticated (for example, aesthetic and philosophical interests), give shape to our goals, and the actions, practices, and processes that give rise to control are directed toward them. To have *goals* or *aims* is at the core of what is to be a controller.

Second, contrary to what the definition requires, in order for *A* to have control over *B* it does not suffice to have the mere capacity or disposition to drive *B* into a certain state. The reason is that *A* might have the capacity to drive *B* into whichever state *A* wants and yet refuse to do so. In that case, although *A* has the disposition to control *B*, A does not de facto control *B*.[30]

Third, what does "drive" mean? That is, what is the nature of the control relation? In most cases, the relation is *causal*: *A* controls *B* by causing *B* to

---

[29] Ascriptions of control are made about all sorts of things: cars, emotions, persons, animals, the volume, passports, the crime rate. Plausibly, what we mean by such ascriptions is that we control behaviors, events, or states related to them.

[30] This is, I believe, a potential problem for Coffman's definition of control (Coffman 2009). Coffman (see point (2.2) in the introductory section of my essay) cashes out control in terms of being free to perform certain actions. However, being free to φ is compatible with deciding not to φ. Specifically, (2.2) entails that if one has the choice to φ but decides not to, one has control over the event that the action of φ-ing aimed to control. Yet one may not have de facto control over the event precisely because one has not φ-ed. As regards (2.3), the epistemic conditions it includes are, as Levy (2011, chap. 5) acknowledges, so demanding that agents rarely satisfy them (so most of their actions are by luck). In my opinion, (2.3) does not match our ordinary notion of control, at least the kind of control that matches the *ordinary* way we think about luck (which might be different from the way philosophers think about luck in the free will debate).

be in a certain state.[31] These qualifications of Dennett's definition allow us to distinguish a specific sense of the notion of control, which I will call *effective control*:

> *Effective Control*: A has effective control over B if and only if (i) it is A's aim that B is in a certain state S, (ii) A has a disposition to cause B to be in S, and (iii) B is in S because of A's disposition to cause B to be in S.

A driver safely driving his car has effective control over his car because (i) he has the aim and (ii) the disposition to maintain or modify the trajectory of the car, its speed, and so on (a disposition that must be stable and integrated with the other driver's dispositions), and (iii) those parameters are so because of the driver's disposition. In general, all instances of control that involve causal influence of the controller on the controllee are instances of effective control.

Consider now the following possible situation. A has the relevant aim and the disposition to drive B into a certain state; B is already in the state A wants B to be; A has done nothing to drive B into that state (that is, B's being in that state is not because of A's disposition). Since condition (iii) is not satisfied, A does not have effective control over B. Can A still have control over B? As the following examples show, the answer is positive:

1. NASA sends an astronaut to the moon. The launch, trajectory, speed, and landing of the spacecraft have been carefully planned by the NASA engineers in such a way that, if all the parameters are as expected, the spacecraft will automatically land on the moon without the need for the astronaut to intervene. If, however, an unforeseen event were to have changed some of the relevant parameters, the astronaut would correct them. When everything goes as planned, the astronaut exerts no causal influence on them. We can still claim, however, that the astronaut, when checking the control panel, has control over the spacecraft.

2. The main concern of doctors is to keep patients healthy. When patients are ill, doctors apply the most adequate treatment and constrain patients to adopt healthy habits. In this way, they exert causal influence on the parameters that determine their patients' state of health. But when patients are healthy, doctors just keep an eye on them and eventually run tests to assess their current state of health, so that if something turns out to be abnormal, they can be in a position to provide proper medical care. In this latter case, doctors have

---

[31] Some uses of the term "control," however, may not allow for causation. Suppose that it is acceptable to say of some mental event that it controls a physical event. Some philosophers might not be willing to qualify the relation between these events as causal (perhaps they would prefer to qualify it as a relation of *determination*). The definition can be tweaked accordingly.

no causal influence on the parameters that determine the patients' state of health, but it still makes sense to say that they have control over them.

What astronauts and doctors have in common is that they *monitor* relevant parameters that, respectively, determine the spacecraft's trajectory and speed and the patients' state of health. Even if causal influence is not sustained, they can have control over them in this way. I call this form of control *tracking control*:

Tracking Control: *A* has tracking control over *B* if and only if *A* monitors *B*.[32]

Monitoring has two components. When *A* monitors *B*, *A* keeps track of, registers, or gathers information about *B*. This is the *epistemic* or *informational component* of monitoring. In addition, the information that *A* registers about *B* enables *A* to initiate, stop, or continue some performance or action that contributes to the achievement of the relevant goal; in a sense, the information compiled disposes or puts *A* in a position to perform goal-directed actions. This is the *dispositional component* of monitoring.

When only the first component is in place, *A* carries out a *merely informational monitoring* of *B*. This is the case of an eventual eavesdropper, who just wants to find out what other people are saying. Ascriptions of control might be true when *A* carries out merely informational monitoring of *B*. This would apply to such ascriptions as "The eavesdropper controls the conversation." Nevertheless, for the most part, monitoring is not merely informational but also *dispositional*, as when a spy eavesdrops in order to get crucial information that could stop an ongoing conflict.

What do we mean when we say that *A* controls *B* in an ordinary sense? When used in ordinary contexts, the term "control" may refer to: (1) effective control, (2) tracking control through merely informational monitoring, (3) tracking control through dispositional monitoring, and (4) a combination of effective and tracking control (plausibly through dispositional monitoring). Case (2) is exemplified by the eventual eavesdropper case above. The following exemplifies (1), (3), and (4): the ascription "The doctor controls the patient's infection" is true if a doctor runs tests to determine the

---

[32] Something like monitoring is what Riggs (2009) seems to have in mind when he proposes that luck arises only if the agent does not successfully *exploit* the relevant event for some purpose. He exemplifies the point with a case in which someone exploits an eclipse in his favor (namely, to survive). Riggs argues that knowing that there will be an eclipse and putting that knowledge into action prevents the eclipse from being lucky for that person, as a simplistic lack of control account of luck would entail. Nevertheless, Riggs does not seem to consider the exploitation of an event in one's favor as a form of control: "The eclipse was not a matter of luck for [that person] because, *though it was out of his control*, he nonetheless exploited its occurrence to procure his survival" (Riggs 2009, 218; emphasis added). Monitoring, I argue, *is* a form of control.

cause of the infection—(3)—if she administers antibiotics to the patient—(1)—and if she does both things—(4). It is worth noting that in most cases tracking and effective control go hand in hand, especially when control follows feedback or feedforward schemas.

The question this section has aimed to answer is the following: In which sense are we at risk with respect to significant events beyond our control and how does this bear on luck? In other words, how should we understand the term "control" in AR and consequently the relevant lack of control condition for luck (given that, as I have argued, luck arises just in case there is agent-relative risk)? The term "control" in AR should be read in the ordinary sense. This means that, depending on what form of control is salient in the context, one might be at risk with respect to an event either when one lacks effective control, or tracking control, or both.

Nevertheless, although the notion of control in AR is the ordinary one (so that context makes salient the relevant form of control in each case), I am inclined to exclude merely informational monitoring as a proper form of control, that is, as a form of control that is able to put an agent in a safe position with respect to an eventuality. Consider again the eavesdropper case. Suppose that the conversation the eavesdropper is controlling gives him information about a potential risk for him. If he is monitoring the conversation in a merely informational manner such that he does not thereby acquire a disposition to act in a way that would put him in a safe position, he is still at risk with respect to the potential occurrence of the event. In brief, the only kind of tracking control that may exclude agent-relative risk is dispositional tracking control. *The lack of control condition in* AR *should be read accordingly.*

We can now state in which way Vincent (in *Buried Treasure*) lacks control over his discovery. The ascription "Vincent controls his discovery" is false not in virtue of a lack of effective control (arguably, Vincent exerts some degree of effective control when digging) but in virtue of a lack of tracking control. To compare: things would have been different with a metal detector, as he could have monitored the location of the treasure and have thereby acquired a disposition to exert proper effective control over his discovery. In that case, his discovery would not have been by luck.

7. The Lack of Control Account of Luck and Its Counterexamples

My view on luck, in a nutshell, is as follows: An event E is lucky or fortunate for S if and only if E occurs and S is at risk with respect to E. S is at risk with respect to E if and only if (i) S has an interest N, (ii) if E were to occur, it would have some objectively positive or negative effect on N, and (iii) S lacks control over E. Condition (iii) must be understood as follows: S lacks control over E if and only if S lacks either effective control, dispositional tracking control, or both. Context and the type of event that E is make salient the form of control that is relevant to assess (iii) in each case. In addition, there are two types of luck or fortune: type A and type B, whose paradigmatic examples are, respectively, winning a fair lottery and winning a lottery that, unbeknownst to one, has been rigged in one's favor. An event E is A-lucky or A-fortunate for S if and only E occurs but was at risk of not occurring and S is at risk with respect to E. E is at risk of not occurring if and only if it would fail to occur in a large enough proportion of close possible worlds. An event E is B-lucky or B-fortunate for S if and only if E occurs, E was *not* at risk of not occurring, and S is at risk with respect to E. In sum, luck essentially arises out of lack of control, but not taking into account the modal profile of lucky events means overlooking the important difference between the A-type and the B-type of luck.

Lackey (2008) has offered counterexamples to the lack of control account of luck (my view is a version of it), both to the claim that lack of control over a significant event suffices for the event to be lucky and to the claim that being a lucky event entails lack of control over it. Against the sufficiency claim, Lackey argues that one's neighbor's playing a computer game right now, one's cat sleeping this afternoon, or a chef's making eggplant parmesan in Florence today are nonlucky events over which one lacks control. This set of counterexamples can be easily ruled out with the significance condition that is embedded in AR: one is not at risk with respect to those events, because one does not have an interest N such that the events have some objectively positive or negative effect on N. To put it briefly, the proposed counterexamples are cases of nonsignificant events.

Lackey proposes another set of counterexamples to the sufficiency claim that do involve significant events: a daughter being picked up from school by her mother, a daughter being properly cared for by her father, and the sun rising every day. As regards the first two events, Lackey (2008, 258) admits that "there is a sense in which both of the events discussed above are lucky: [the child's father] is lucky that he has the sort of wife whom he can depend on to pick up their children, and [the daughter] is lucky that she has a father who takes proper care of her," but she quickly withdraws the claim, arguing that this sense of luck would make us deem too many events as being lucky, and it clearly differs from the sense of luck of the type of events proponents of the lack of control account are interested in (for example, winning a fair lottery). As regards sunrises, Lackey thinks that it is clear that we lack of control over them and that they are not lucky for us.

I have two comments on this set of counterexamples. First, it is not obvious that there is no control involved. As regards the first case (a daughter being picked up from school by her mother), it is not clear that the child's father does not control the fact that his daughter is being picked up from school safely if, say, he can phone (monitor) his wife at any time to know whether she is on her way to pick her up. When the father performs that or similar actions, he has tracking control. As regards the second case (a daughter being properly cared for by her father), it is not clear that the daughter has no control over the events that make her be properly cared for if, say, the child knows of the existence of social services and knows how to call for aid to remedy the eventual carelessness of her father. When the daughter performs that or similar actions, she can monitor relevant parameters of her own situation (tracking control) and call for aid (effective control).[33] Unfortunately, Lackey's cases are underdescribed. Finally, it is clear that, although we have no causal influence on the sun and hence no effective control over it, we do monitor sunrises in a dispositional way, that is, in a way that *disposes* us to perform goal-directed actions. When one keeps track of the time the sun rises, one can count on it, for example, to wake up and go to work. The same applies to many other nomic necessities: when one is able to monitor an event that occurs as a matter of natural law, one is able to count on the occurrence of that event safely. In sum, my account of the notion of control explains why all these alleged counterexamples are not cases of luck.

Second, suppose that it turns out that there is lack of control in Lackey's examples. Then, the sense in which the relevant events would be lucky would be the sense in which B-cases are cases of luck: although there is no risk that the events will fail to occur, the agents in question are at risk with respect to them precisely because of their lack of any form of control. Lackey is right in thinking that the sense of luck that arises in this kind of cases, which is clearly different from the sense of luck involved in, for example, winning a fair lottery, make us deem too many events as being lucky. It is no objection, however, to the lack of control account of luck I am presenting here that it leads to a proliferation of lucky events. As I have argued, luck is a more pervasive phenomenon than many commentators think, and this is reflected by our ordinary ascriptions of luck. My account of luck in terms of risk explains in a motivated way why it is so pervasive.

On the other hand, Lackey also charges against the necessity claim (the claim that if an event is lucky for one, one lacks control over it). She proposes the ingenious *Demolition Worker* counterexample:

> Ramona is a demolition worker about to press a button that will blow up an old abandoned warehouse, thereby completing a project that she and her co-workers have been working on for several weeks. Unbeknownst to her, however, a mouse had chewed through the relevant wires in the construction office an hour earlier, severing the connection between the button and the explosives. But as Ramona is about to press the button, her co-worker hangs his jacket on a nail in the precise location of the severed wires, which radically deviates from his usual routine of hanging his clothes in the office closet. As it happens, the hanger on which the jacket is hanging is made of metal, and it enables the electrical current to pass through the damaged wires just as Ramona presses the button and demolishes the warehouse. (Lackey 2008, 258)

---

[33] Calling for aid seems a form of effective control, as the daughter's words are the origin of the causal chain of events that leads to the relevant aid.

Lackey claims that the explosion is both under Ramona's control and by luck. However, does merely pressing a button suffice for having control over an explosion, at least in this case? Hardly, since an important feature of the case, as Lackey introduces it, is that Ramona is one of the persons who have been working for weeks on the design of the controlled explosion, which means that the extent to which Ramona should have control over the explosion encompasses not only the mere production of the explosion by pressing the button (effective control) but also the monitoring of the explosion system (tracking control). In particular, not having properly checked the relevant wires before the explosion or having failed to foresee the presence of rodents or other problematic animals are things for which we can blame Ramona and the rest of the co-workers responsible for the design of the alleged controlled explosion. If they had monitored these things adequately, they would have been in a position to take proper action and to demolish the warehouse in a way that we would not classify as lucky. In conclusion, although Ramona has effective control over the explosion, she lacks tracking control over it, and in the context provided by Lackey both forms of control are salient. Therefore, the explosion occurs by luck, just as the lack of control account predicts.

8. Conclusions

What is luck? In this essay, I have argued that the concept to which the terms "luck" and "fortune" refer in ordinary discourse can be adequately defined in terms of risk. In particular, I have argued that luck arises just in case an agent is at risk with respect to an event, which means, in turn, that luck arises just in case an agent lacks control over a significant event. I have also argued that in order to understand properly what luck is, we must take into account not only the risk at which an agent is with respect to lucky events but also the risk that lucky events had of not occurring, which means, in turn, that we must take into account their modal profile. Only in this way can we fully appreciate the distinction between two intuitively different forms of luck: the kind of luck that is instantiated when one wins a fair lottery, and the kind of luck that is instantiated when one wins a lottery that, unbeknownst to one, someone has firmly decided to rig in one's favor. Finally, I have given an account of the notion of control that explains why my version of the lack of control account of luck steers clear of several counterexamples.

References

Balaguer, Mark. 2010. *Free Will as an Open Scientific Problem*. Cambridge, Mass.: MIT Press.
Ballantyne, Nathan. 2012. "Luck and Interests." *Synthese* 185, no. 3:319–34.

Baumann, Peter. 2012. "No Luck with Knowledge? On a Dogma of Epistemology." *Philosophy and Phenomenological Research*, DOI: 10.1111/j.1933-1592.2012.00622.

Broncano-Berrocal, Fernando. 2014. "Is Safety in Danger?" *Philosophia* 42, no. 1:63–81.

Coffman, E. J. 2007. "Thinking About Luck." *Synthese* 158, no. 3:385–98.

———. 2009. "Does Luck Exclude Control?" *Australasian Journal of Philosophy* 87, no. 3:499–504.

Dennett, Daniel C. 1984. *Elbow Room: The Varieties of Free Will Worth Wanting.* Oxford: Oxford University Press.

Lackey, Jennifer. 2006. "Pritchard's Epistemic Luck." *Philosophical Quarterly* 56, no. 223:284–89.

———. 2008. "What Luck Is Not." *Australasian Journal of Philosophy* 86, no. 2:255–67.

Latus, Andrew. 2003. "Constitutive Luck." *Metaphilosophy* 34, no. 4:460–75.

Levy, Neil. 2009. "What, and Where, Luck Is: A Response to Jennifer Lackey." *Australasian Journal of Philosophy* 87, no. 3:489–97.

———. 2011. *Hard Luck: How Luck Undermines Free Will and Moral Responsibility.* Oxford: Oxford University Press.

McKinnon, Rachel. 2013. "Getting Luck Properly Under Control." *Metaphilosophy* 44, no. 4:496–511.

Pritchard, Duncan. 2005. *Epistemic Luck.* Oxford: Oxford University Press.

———. 2015. "The Modal Account of Luck." Included in this collection.

———. Manuscript. "Risk."

Rescher, Nicholas. 1995. *Luck: The Brilliant Randomness of Everyday Life.* New York: Farrar, Straus and Giroux.

Riggs, Wayne D. 2009. "Luck, Knowledge, and Control." In *Epistemic Value*, edited by Adrian Haddock, Alan Millar, and Duncan Pritchard, 204–21. Oxford: Oxford University Press.

Steglich-Petersen, Asbjørn. 2010. "Luck as an Epistemic Notion." *Synthese* 176, no. 3:361–77.

Unger, Peter. 1968. "An Analysis of Factual Knowledge." *Journal of Philosophy* 65, no. 6:157–70.

Williamson, Timothy. 2009. "Probability and Danger." *The Amherst Lecture in Philosophy 4*, 1–35. Available at http://www.amherstlecture.org/williamson2009/

CHAPTER 2

STROKES OF LUCK

E. J. COFFMAN

A wide range of debates across such areas as epistemology, ethics, philosophy of action, philosophy of law, and political philosophy center on luck-
involving claims like these:

- If you know, then it's not lucky that you believe accurately.
- If it was lucky that you acted as you did, then you did not freely so
  act.
- If you and I behave in the same way but my conduct has worse results
  than yours through sheer bad luck, I am no more blameworthy than
  you are for so behaving.
- We should redistribute resources so as to enhance the prospects of
  those who, through sheer bad luck, are among our worst off.
- We can properly punish successful criminal attempts more severely
  than ones that fail only by luck.

Such claims can impart a strong sense that "[t]he concept of luck plays a
crucial role in many philosophical discussions" (Lackey 2008, 255). Under
this impression, a number of philosophers have recently begun developing and evaluating new, and unusually rigorous, accounts of luck.[1] This
research program promises dividends whether or not the concept of luck
really is as important as claims like those above suggest. If the concept
actually does play a "crucial role" in some or other of the indicated debates,
then working toward its correct analysis will advance those debates rather

[1] Noteworthy contributions include Reseller 1995, Latus 2003, Pritchard 2005, Riggs 2007
and 2009, Lackey 2008, and Levy 2011; other contributions include Coffman 2007 and 2009.

The Philosophy of Luck, First Edition. Edited by Duncan Pritchard and Lee John Whittington.
Chapters © 2014 Metaphilosophy LLC and John Wiley & Sons Ltd, except for "Luck as Risk
and the Lack of Control Account of Luck" © 2015 Metaphilosophy LLC and John Wiley &
Sons Ltd. Book compilation © 2015 Metaphilosophy LLC and John Wiley & Sons Ltd.
Published 2015 by John Wiley & Sons Ltd.

directly, by progressively clarifying claims that drive them. But perhaps initial appearances mislead: maybe those discussions don't really revolve around *luck*, but instead revolve around some similar, more or less closely related notion(s). If so, then homing in on the correct analysis of luck should help us recognize that our focus on it is misplaced, which should in turn lead to beneficial clarification of claims like those above.

This essay aims to reorient current theorizing about luck as an aid to our discerning the concept's true philosophical significance. After introducing the literature's leading theories of luck, I present and defend counterexamples to each of them (sections 1 and 2). Next, I argue that recent luck theorists' main target of analysis—the concept of an event's being *lucky for* a subject—is actually parasitic on the more fundamental notion of an event's being a *stroke of luck for* a subject, which thesis will serve as at least a partial diagnosis of the leading theories' failure (sections 3 and 4). I then develop an analysis *of strokes of luck* that utilizes insights from the recent luck literature (section 5). Finally, having set out a comprehensive new analysis of luck—what I call the *Enriched Strokes Account* of lucky events (section 6)—I revisit the initial counterexamples to the literature's leading theories and argue that the Enriched Strokes Account properly handles all of them (section 7).[2]

Before diving in, let me flag an important assumption and describe how I use some important terms in this essay. Following other contributors to the recent luck literature, I assume that the luck relation(s) can relate (a) individuals for whom things can go better or worse to (b) events proper as well as obtaining states of affairs (or facts). What I call "events proper" are concrete-object-like entities that have spatiotemporal locations and are denoted by perfect gerundial nominals—for example, "Ann's catching of the ball," "the shark's biting of Bob." States of affairs, by contrast, are proposition-like entities that obey Boolean principles and are denoted by imperfect gerundial nominals—for example, "Ann's catching the ball," "the shark's biting Bob."[3] Accordingly (and in line with other luck theorists), I use "event" in a broad sense that covers events proper as well as states of affairs. I use "happen" in a broad sense that covers both *occurrence* (for events) and *coming to obtain* (for states of affairs). And I use "do" in a broad sense that covers both *performance* (for events) and *actualization* (for states of affairs).

1. Three Leading Theories of Luck

Say that possible world W_1 is *close to* world W_2 before time t iff W_1 differs no more than slightly from W_2 up to (but not including) t. With this

---

[2] In Coffman forthcoming, I bring the Enriched Strokes Account of lucky events to bear on some central debates in epistemology and philosophy of action.

[3] The last two sentences owe much to Bennett 1988 as well as Paul and Hall 2013.

stipulative definition in hand, we can state the literature's three leading accounts of luck as follows:

The Modal Account: Event E is at t (un)lucky for subject S = df. (E happens at t and) (i) E is in some respect good (bad) for S and (ii) E doesn't happen around t in a wide class of possible worlds close to the actual world before t.[4]

The Control Account: E is at t (un)lucky for S = df. (i) E is in some respect good (bad) for S, (ii) S hasn't successfully exploited E for some purpose, and (iii) E isn't something S did intentionally.[5]

The Mixed Account: E is at t (un)lucky for S = df. (i) E is in some respect good (bad) for S, (ii) E doesn't happen around t in a wide class of worlds close to the actual world before t, and (iii) E isn't something S did intentionally.[6]

A few remarks about each account's condition (i), and the Modal and Mixed Accounts' condition (ii). Condition (i) seems the best way to understand the so-called significance condition on luck.[7] Since an event can be good for you in one respect but bad for you in another, accounts of luck that utilize (i) correctly allow that an event can be both lucky *and* unlucky for you (cf. Ballantyne 2012, 331). For example, your lottery win may be good luck in that it enables you to retire early, but bad luck in that it makes you a salient target for extortion.

For statements of the "chanciness" condition on luck that resemble the Modal and Mixed Accounts' condition (ii), see Pritchard 2005, Coffman 2007, and Levy 2011. It's important to state (ii) with *"around* t" instead of *"at* t." If (ii) is stated with "at," each account's right-to-left conditional would be vulnerable to this sort of counterexample: Under perfectly normal conditions, you (automatically, non-intentionally) inhale wholesome air at t. Inhaling wholesome air is good for you, and it doesn't happen at t in a wide class of worlds close to the actual world before t (in most such worlds, you're either exhaling or "idle" at t). If (ii) is stated with "at," each account's right-to-left conditional implies incorrectly that you are at t lucky to be inhaling wholesome air. With "around," each account's right-to-left conditional avoids this implication (assuming that you inhale

---

[4] For careful development of an influential version of the Modal Account, see Pritchard 2005. (Henceforth, I'll typically suppress the parenthetical "occurrence" clause, which should be assumed in all the accounts of luck discussed below.)

[5] This nuanced version of the Control Account is due to Riggs 2009. For simpler versions that don't include anything like condition (ii), see Nagel 1976, Zimmerman 1987, Statman 1991, and Greco 1995. We'll consider Riggs's rationale for adding condition (ii) below.

[6] For recent versions of the Mixed Account, see Coffman 2007, Riggs 2007, and Levy 2011.

[7] For the best available discussion of the "significance" (or, perhaps better, "value") condition, see Ballantyne 2012.

wholesome air *around* t in the vast majority of worlds close to the actual world before t).[8]

Moreover, (ii) allows the Modal and Mixed Accounts to countenance lucky events in settings where causal determinism obtains.[9] Numerous cases illustrate this possibility (cf. Pritchard 2005, Coffman 2007, Levy 2011). Winning the lottery in a deterministic world is lucky for you, notwithstanding the fact that your win was necessitated by prior events and the laws of nature. For another example, suppose you live in a deterministic world where your life depends on a certain sphere's remaining perfectly balanced on the tip of a particular cone throughout some temporal interval.[10] We can fill in the details so as to elicit the intuition that you are lucky the sphere remains perfectly balanced on the tip of the cone throughout that interval, notwithstanding the fact that the sphere's remaining so balanced was necessitated by prior events and the laws of nature.

Each of our leading accounts issues correct verdicts about certain clear cases of luck like these:

> *Good Lottery:* You habitually play numbers corresponding to your own birthday in the state lottery. On this occasion, however, you seriously contemplate playing numbers corresponding to your mother's birthday. In the end, you stick with standard practice and play your own numbers. Lo and behold, you win!
> *Bad Lottery:* You live in a corrupt state where citizens are forced to play in a lottery whose "winners" lose their life savings to the governor. As before, you vacillate between playing your mother's birthday numbers and your own birthday numbers. In the end, you stick with your own numbers. Lo and behold, you "win"!

In each example, your lottery win isn't something you did intentionally; you haven't yet exploited your win for some purpose; and, in a wide class of worlds close to the actual world before the time at which you win, you don't win around then.[11] So, provided that your win in *Good Lottery* is good for you—and that your "win" in *Bad Lottery* is bad for you—each of our leading accounts entails that your win is (un)lucky for you. Unfortunately, as we're about to see, each account is also vulnerable to successful counterexample.

---

[8] Thanks to Georgi Gardiner for helping me get clearer on these issues.

[9] "Causal determinism" here denotes the thesis that "there is at any instant exactly one physically possible future" (van Inwagen 1983, 3).

[10] This case is inspired by one due to Williamson 2000, 123.

[11] Consider various small changes we could make in the actual world before the time when you won (I assume you played "062976"): "6" is the penultimate number selected; "5" is the penultimate number selected; and so on. If things had been slightly different in one of these ways before the time when you won, you wouldn't have won around then. So, in a wide class of worlds close to the actual world before the time when you won, you don't win around then.

2. Counterexamples to the Leading Theories

Start with the Modal Account. Over the next few paragraphs, I'll defeat Levy's (2011, 20-22) attempted defense of condition (ii)'s necessity for luck from the following counterexample due to Lackey (2008, 261, paraphrased):

> *Buried Treasure:* Sophie buries her treasure at the one spot where rose bushes can grow on the northwest corner of her island. Sophie was set on burying her treasure on the island's northwest corner in a spot that supports rose bushes: that's her favorite part of the island, and roses are her favorite flowers. All this is unbeknownst to Vincent, who shows up one month later at the exact same spot where Sophie buried her treasure. Vincent's reasons for digging up that spot are different from, and completely unrelated to, Sophie's: Vincent is set on planting a rose bush in his mother's memory on that part of the island. As Vincent digs, he's shocked to find buried treasure.

Finding Sophie's treasure when he does seems lucky for Vincent. But given the details, Vincent finds Sophie's treasure around that time in the vast majority of possible worlds close to the actual world before he finds it. *Buried Treasure* looks to be a counterexample to (ii)'s necessity for luck.

Levy (2011, 20-22) attempts to defend (ii) from cases like *Buried Treasure*. We can understand Levy as trying both to undercut and to rebut the judgment that Vincent is lucky to find Sophie's treasure. As for the undercutter, Levy suggests that the objector's judgment that Vincent is lucky to find the treasure stems from the fact that the discovery seems lucky to Vincent. But since E's seeming lucky to S is consistent with E's not actually being lucky for S, the objector's reason for reckoning Vincent's discovery lucky doesn't justify that judgment.

Levy's attempt to rebut the judgment that Vincent is lucky to find Sophie's treasure takes off from the following case:

> *Buried Treasure**: Unbeknownst to Vincent, Sophie buried the treasure in the spot at which he found it because Vincent's eccentric great-uncle wanted him to have the riches (perhaps Sophie was unaware of the plan; perhaps Vincent's great-uncle is a neuroscientist with the power to implant in Sophie a love of roses, knowing it will lead her to bury her treasure in the one spot where he knows Vincent will dig). In that case, it will seem to Vincent very lucky that there was treasure in the precise spot at which he dug, but luck has nothing to do with it; his finding the treasure was planned. Indeed, that might be precisely how Vincent's great-uncle thinks of his plan: he has, as he might say, "left nothing to chance." (Levy 2011, 21)

According to Levy, if Vincent knew all the details of *Buried Treasure* and how that case relates to *Buried Treasure** (in which Vincent's discovery clearly isn't lucky), Vincent should think he really isn't lucky in *Buried Treasure* after all. The fact that Vincent should think this in light of such knowledge is evidence that he isn't lucky in *Buried Treasure*.

Reply: We can sidestep Levy's attempted undercutter by deleting from *Buried Treasure* the unnecessary detail that Vincent is surprised to find Sophie's treasure, and stipulating instead that Vincent was confident he'd make such a discovery despite having no evidence whatsoever for this proposition. The discovery no longer seems lucky *to* Vincent, but it still seems lucky *for* him (cf. Steglich-Petersen 2010, 365ff.). Thus, the basis for our intuition that the discovery is lucky for Vincent can't just be that it seems lucky to him. As for the attempted rebutter, note that we must give its occurrences of "should" an "epistemic" reading.[12] But since there's a highly relevant difference between *Buried Treasure* and *Buried Treasure** (viz., the presence of the great-uncle's planning/design), it's not at all clear that Vincent should (epistemically) infer, from the details of *Buried Treasure* and its relation to *Buried Treasure**, that he's not lucky in the former. Indeed, that seems like a pretty weak analogical inference, given the highly relevant difference between the two cases in terms of overall planning/ design (cf. Lackey 2008, 262–63).[13]

Having defended Lackey's counterexample to the Modal Account's left-to-right conditional, let's consider its right-to-left conditional. Contrary to this portion of the Modal Account, a morally significant yet "modally fragile" choice needn't be a lucky occurrence (cf. Latus 2003, Coffman 2007, Lackey 2008). Suppose that you choose at t to make a substantial donation to Oxfam, where there was just before t a large chance you would not so choose at t (you had fairly strong "self-interested" reasons to omit such a choice, and so on). The Modal Account entails that you're lucky you chose to donate. But suppose your relevant reasons inclined you at least somewhat toward making the donation, and that the choice and subsequent donation were made intentionally, freely, and knowingly.[14] Then your choosing to donate wasn't—or at least, it needn't have been—*lucky* for you, notwithstanding the fact that your choice was both modally fragile and good for you.

---

[12] If we don't give Levy's "should" an "epistemic" reading, then we should (epistemically) reject his assumption that a subject's being such that the subject should self-ascribe luck is evidence that she is in fact lucky.

[13] Some theorists have inferred, from the success of counterexamples like *Buried Treasure*, that the kind of chance or improbability related to luck is not *metaphysical* but instead *epistemic*. One prominent recent proponent of this type of view is Steglich-Petersen (2010, 369; cf. Rescher 1995, 28–34). See my footnote 24 for brief critical remarks on Steglich-Petersen's proposal, which I discuss at greater length in section 1.4 of Coffman forthcoming.

[14] Remember that "modally fragile" doesn't entail "undetermined." My claims here are thus neutral on the issue whether undetermined events can be acts done intentionally, freely, and knowingly.

We turn now to the nuanced version of the Control Account recently developed by Riggs:

E is at t (un)lucky for S = df. (i) E is in some respect good (bad) for S, (ii) S hasn't successfully exploited E for some purpose, and (iii) E isn't something S did intentionally. (Riggs 2009, 220)

Riggs's addition of (ii) to simpler versions of the Control Account immunizes his analysis against cases involving significant states of affairs that you didn't bring about intentionally but are nevertheless *not* lucky for you, given their "modal stability." Despite the fact that you didn't intentionally bring about the sun's rising this morning or the continued functioning of the electricity in the room you're now in (assuming you're currently inside, and so on), neither of these states of affairs counts as *lucky* for you. While such examples refute simpler species of the Control Account, they don't refute Riggs's version (assuming you've successfully exploited the indicated states of affairs for some purpose or other, which you presumably have). Moreover, Riggs's endorsement of (ii) is independently motivated, for he adds it in light of intuitions about this fascinating case:

African Expedition: Smith plans an expedition into the wilds of Africa where certain tribes of Africans with exotic customs were known to live. Smith is constrained … to make this trip during a particular month of a particular year. He proposes the trip to his fellow adventurer Jones, including the specific times that he means to travel. Jones agrees to tag along. As it happens, the particular tribe that lives in the area that Smith and Jones visit has a custom of sacrificing people from outside the tribe on the equinoxes of the year. The autumnal equinox happens to fall during the time that Smith and Jones are in the area, so they are captured and held until that day so that they can be sacrificed. … As the tribesmen approach to kill them …, there is a total eclipse of the sun. The members of this tribe always take such exotic natural occurrences to signal the anger of their gods at them for whatever they happen to be doing at the moment. Consequently, they set their captives free. … Smith says to Jones, "That solar eclipse was an amazing stroke of good luck!" Jones replies, "Don't be absurd! There was nothing lucky about it. I knew all along that these people would likely try to sacrifice us on the equinox if we were captured, but I also knew that there was a total eclipse of the sun due on that very day, and that this tribe would react to that event by letting us go. Did you really think I would be stupid enough to fall into such a situation without having a plan to extricate myself?" (Riggs 2009, 216)

According to Riggs, "Smith was lucky that the eclipse happened, and Jones was not" (2009, 217). Obviously, neither adventurer intentionally brought about the eclipse, so we'll need something other than (iii) to capture our sense that the eclipse was lucky for Smith but not for Jones. Here's where (ii) comes in. Since only *Jones* successfully exploited the eclipse for some purpose—that is, only Jones "planned a course of action that assumed that [the eclipse] would occur," which secured his survival (2009, 218)—(ii) and (iii) together imply that the eclipse was lucky only for *Smith*.

Alas, Riggs's Control Account's right-hand side doesn't suffice for luck. Suppose that Katelyn lives and works in an underground facility that is, *unbeknownst to her*, solar powered. This morning's sunrise was good for Katelyn (it kept her facility running), not intentionally brought about by her, and not successfully exploited by her for some purpose (having been underground for so long, Katelyn has become oblivious to sunrises, and thus doesn't plan any courses of action assuming that the sun will rise). Riggs's Control Account implies, incorrectly, that Katelyn was lucky the sun rose this morning.

As for the left-to-right conditional of Riggs's Control Account, cases like the following show that (ii) isn't required for luck:

> *Kidnapping*: My son Zachary is kidnapped. The kidnappers communicate to me that I can have Zachary back as soon as I tell them that I can pay a huge ransom. Once I tell them this, I'll have twenty-four hours to deliver the payment; if I don't get the money to them before the deadline, they'll recapture Zachary and increase the ransom. An informant I know to be extremely reliable tells me that tonight's lottery will be rigged in my favor. According to my informant, the lottery officials have heard about my family's plight and want to help us out. So I form a justified belief that I'll win tonight's lottery. Now, on this occasion, my reliable informant is mistaken: tonight's lottery won't be rigged in my favor. As it happens, though, I'm going to win it fair and square! So I have a justified and *true* belief that I'm going to win tonight's lottery. I phone the kidnappers to tell them that I can pay the ransom. They release Zachary, reminding me that if they don't have the money in twenty-four hours, they'll recapture him and increase the ransom. All along, I'm thinking: "No problem—I'll have my lottery winnings in hand with plenty of time to spare!" I win the large, fair lottery that evening.

My lottery win was a huge stroke of good luck for me. And it's a fact that I successfully exploited for the purpose of getting Zachary back: I planned a course of action assuming I would win, which resulted in Zachary's safe return (compare Jones's relation to the eclipse in *African Expedition*). Moral: An event can be lucky for you *even if you* successfully exploit it for some purpose.

As for condition (iii) of the Control Account, cases like the following show that it's not necessary for luck:[15]

---

[15] For two additional structurally similar examples, see Lackey's "Demolition Worker" and "Derek" cases (Lackey 2008, 259–60).

Distracted Driver: Our department meeting just happens to end early for once. As a result, I'm early to pick up my son Evan from school. Upon arriving, I spot him playing in the street. A car, whose distracted driver is texting on her cell phone, speeds toward Evan. But I'm in a position to push Evan out of the car's path, and I of course do so. Hugging Evan tightly moments later, I say: "I'm so lucky you're safe!"

If your intuitions follow mine here, this self-ascription of luck will strike you as true *on balance* (more on this qualification in a moment). The seeming truth (on balance) of that self-ascription can be bolstered by the even stronger sense that the self-ascription is fully appropriate, combined with the "general presumption that, where speakers are not basing their claims on some false beliefs they have about underlying matters of fact, how they naturally and appropriately describe a situation ... will be a true description" (DeRose 2009, 51). I conclude, then, that I was (at the indicated time) lucky that Evan was safe. Note, finally, that we can understand this case so that Evan's safety is something I brought about intentionally (by pushing him out of the car's path). Moral: An event can be lucky for you *even if* it's something you did intentionally.

Now, why the "on balance" qualification? Because I expect that some readers will feel some inclination to deny that I am (at the relevant time) *lucky* that Evan is safe (cf. Levy 2011, 22–23). I can understand such doubt, having recently harbored it myself (cf. Coffman 2009, 503). But I now view such doubt as a mistaken response to a different, though closely related, fact—viz., that it's not a *stroke of good luck* for me that Evan is safe. To be sure, Evan is safe *by* a stroke of good luck for me (and for him)—viz., my becoming positioned to save him. But Evan's safety isn't *itself* a stroke of good luck for me (or for him). In the next section, I bolster this error theory by defending the crucial distinction between *lucky events* and *strokes of luck* on which it depends.[16] In section 6, I revisit *Distracted Driver* to show that my overall account of luck yields (what I'm claiming are) the correct verdicts about the case.

We've now seen counterexamples to both directions of both the Modal and the Control Account. As for the Mixed Account, *Buried Treasure* and *Distracted Driver* together show that neither (ii) nor (iii) is necessary for an event's being lucky for you. From the success of counterexamples like *Buried Treasure* and *Distracted Driver* to conditions like (ii) and (iii), Lackey concludes that "the conditions proposed by [the literature's leading accounts of luck] ... fail to capture what is distinctive of, and central

---

[16] The error theory sketched here will serve, mutatis mutandis, to defend Lackey's (2008, 259–60) "Derek" counterexample to the Control Account from the claim that the event she deems lucky for her subject (viz., the making of a free throw) really isn't. Matters are more complicated when it comes to Lackey's "Demolition Worker" case, due to some ambiguity in its details; for pertinent discussion of this case, see section 2.2 of Coffman forthcoming.

to, the concept of luck" (2008, 255). Call this *Lackey's Inference*. In what follows, I'll present and defend a new account of luck on which (ii) and (iii) are requirements on an event's being a *stroke of luck* for a subject, a notion on which the concept *lucky event* is parasitic. If such an approach to luck is correct, then there's a clear sense in which (ii) and (iii) *are* (in Lackey's words) "distinctive of, and central to, the concept of luck" after all. By the end of this essay, then, I believe we'll be in a position to deem Lackey's Inference *hasty*.

3. Lucky Events and Strokes of Luck

Besides their vulnerability to counterexample, the leading theories of luck have another—and, as will become clearer as we proceed, related—defect in common: their proponents have assumed, at least implicitly, that the locution "is (un)lucky for" expresses the most fundamental or basic luck concept. This assumption should strike us as dubious after some reflection on a different locution we frequently use to ascribe luck: "is a stroke of good (bad) luck for." The locutions in question are not equivalent. To see this, extend *Good Lottery* so that the lottery officials are now handing you the lottery's large cash prize. Intuitively, you're lucky to be receiving the prize. But while you are receiving the prize *by* a stroke of good luck (your lottery win), your receiving the prize isn't *itself* a stroke of good luck for you (you're the rightful winner, after all). For another illustration of the point, extend *Bad Lottery* so that the governor is now transferring your life savings into his checking account. Intuitively, you're unlucky to be losing your life savings to the governor. But while you are losing your life savings to the governor *by* a stroke of bad luck (your lottery "win"), the governor's taking your life savings isn't *itself* a stroke of bad luck for you (after all, you're the rightful "winner"). Moral: An event that isn't itself a stroke of good (bad) luck for you may nevertheless be (un)lucky for you.

Keep reflecting on how *is (un) lucky for* might relate to *is a stroke of good (bad) luck for*, and (I predict) you'll eventually feel tempted to deem the latter more fundamental than the former. Whereas there are no promising accounts of *is a stroke of luck for* in terms of *is lucky for* on the horizon, there is a promising analysis of the latter in terms of the former: an event that's *(un) lucky* for you is one that's either *itself a* stroke of good (bad) luck for you or one that's good (bad) for you and due primarily (chiefly, mainly) to some *prior* stroke of good (bad) luck for you. More precisely:

> Event E is at t (un)lucky for subject S = df. (i) E is in some respect good (bad) for S and (ii) there's a stroke of good (bad) luck for S, E*, such that *either* (a) E = E* *or* (b) E* is a primary (chief, main) contributor to E.

Call this the *Strokes Account* of lucky events. This proposal differs crucially from the three leading accounts in that it sees *is lucky for* as disjunctive—as admitting a "direct/indirect" distinction—in a way similar to (for example) *is morally responsible for*: S is morally responsible for an act A iff S is either directly *or* indirectly morally responsible for A, where (i) S is *directly* morally responsible for A iff S is responsible for A but not in virtue of being responsible for something else; and (ii) S is *indirectly* morally responsible for A iff S is responsible for A in virtue of being responsible for something else (cf. Mele 2006, 86–87). Similarly, the Strokes Account's proponent distinguishes between (i) lucky events that *do* inherit their status as such from some other events and (ii) lucky events that *don't* inherit their status as such from any other events (where the latter notion is, the Strokes Account's proponent suggests, commonly expressed in ordinary discourse by "is a stroke of good [bad] luck for" and cognate locutions).[17]

Right away, I must say something about "is a primary (chief, main) contributor to." I intend to use that expression and related ones—for example, "primarily (chiefly, mainly) because"—in their ordinary senses. Such expressions have the same meanings here as they do in familiar bits of discourse like these:[18]

- My having to study *Julius Caesar* in high school is a primary contributor to my present hatred of the work.
- The wind's kicking up when it did was a chief contributor to the fire's spreading into the new timber.
- The driver's reckless behavior was a main contributor to Sara's becoming a widow.

Call the relation that "is a primary (chief, main) contributor to" expresses in the above sentences the *primary contribution relation*; the following claims display its converse:

- I now hate *Julius Caesar* primarily because I had to study the work in high school.
- The fire's spreading to the new timber was due mainly to the wind's kicking up when it did.
- Sara is a widow chiefly because of the driver's reckless behavior.

We can tighten our grip on the primary contribution relation by noting that it's neither necessary nor sufficient for *causal* contribution. First, causes needn't be primary contributors. The Big Bang, nearly all theorists of causation will agree, is a cause of every subsequent event. So the Big Bang is among the causes of every human birth and death. But no human

---

[17] Thanks to Devon Bryson, David Palmer, and Carolina Sartorio for helping me develop this analogy.

[18] The first two examples are adapted from Bennett 1988, 21ff.

birth or death has been due primarily to the Big Bang.[19] Second, primary contributors needn't be causes. Suppose that, at noon, I (Sara's husband) am fatally struck by a reckless driver while crossing the street. Sara becomes a widow, and this is due primarily to the reckless driver's behavior. But while the reckless driver's behavior was surely a cause of my death, it wasn't also a cause of Sara's widowing; the latter is instead a "non-causal consequence" of the driver's behavior.[20] I hasten to add, though, that primary contribution is *like* causal contribution in that both are objective, mind-independent, "out in the world" phenomena (cf. Bennett 1988, 32).

Why "is a primary *contributor to*", rather than "is a primary *cause of*"? The latter would make the Strokes Account too strong. We can imagine that Sara's widowing is unlucky for her, given the way it relates to a certain stroke of bad luck for her—for example, a reckless driver's fatally striking me as I cross the street. But again, the driver's fatally striking me doesn't *cause* Sara's widowing. So, if we replace "is a primary contributor to" with "is a primary cause of," the Strokes Account won't issue the correct verdict that Sara is unlucky to be a widow.

Why "is a *primary* contributor to," rather than merely "contributes to"? The latter would make the Strokes Account too weak.[21] Suppose you narrowly escaped death once as a toddler; your survival then was a stroke of good luck for you. If we go with "contributes to," the Strokes Account's right-to-left conditional will imply that every later significant event to which your survival contributes is lucky for you. Since your survival presumably contributes to *every* later event that involves you, the Strokes Account's right-to-left conditional will imply that *every* later significant "you-involving" event is lucky for you. But such a proliferation of luck would be unacceptable (cf. Levy 2009, 497). Why so? Because the luck analyst's quarry is a phenomenon one paradigm of which is your winning a large fair lottery, and the presence of which can arguably keep true beliefs from constituting knowledge, rational actions from qualifying as free, and so on. Assuming that "lottery win"—type events aren't ubiquitous—and that our ability to gain/retain knowledge, and to act freely, isn't under constant threat from luck—the luck analyst's quarry simply can't be as common a phenomenon as the "contributes to" version of the Strokes Account would have it (cf. Lackey 2008, 258).

---

[19] Cf. Lewis 1986, 23: "[The Big Bang], I take it, is a cause of *every* later event without exception. Then it is a cause of every death. But the Big Bang did not kill anyone."

[20] For starters, since my death arguably doesn't cause Sara's widowing (these happen simultaneously, and possibly at great spatial distance), there seems to be no causal chain connecting the driver's behavior to Sara's widowing (cf. Kim 1974).

[21] The following argument can be adapted to show that replacing "is a primary contributor to" with "is counterfactually sufficient for" would also make the Strokes Account too weak. I leave such adaptation as homework for interested readers. In section 1.3 of Coffman forthcoming, I show that replacing "is a primary contributor to" with "is counterfactually necessary for" would make the Strokes Account too strong.

Happily, the right-to-left conditional of the official statement of the Strokes Account doesn't yield the same unacceptable proliferation of luck. In the case under consideration, for example, we can safely suppose that your survival as a toddler isn't a *primary* contributor to *every* later event in which you're involved. Now it is arguable that, in the envisaged case, every later event that involves you (or at least, every such event of which you're the sole subject) counterfactually depends on your survival as a toddler (if you hadn't survived, that event wouldn't have happened). But counterfactual dependence doesn't suffice for primary contribution. Everything downstream from the Big Bang counterfactually depends on it, but it's not the case that everything downstream from the Big Bang (e.g., your currently reading this sentence) is due primarily to the Big Bang. It's also arguable that, in the imagined case, for every later moment at which you exist, you exist then primarily because you survived as a toddler. But that too is compatible with the claim that your survival as a toddler isn't a *primary* contributor to *every* later event that involves you. Generally speaking, even if X's *existence* at t is due primarily to Y, some X-involving events at t may not be *at all* due to Y.[22] The official statement of the Strokes Account, in terms of the primary contribution relation, seems not to suffer from a problematic over-ascription of luck.

I've now provided a sampling of highly intelligible claims in which the primary contribution relation, or its converse, figures prominently; further clarified the primary contribution relation by distinguishing it from *causal* contribution; and indicated why I've formulated the official version of the Strokes Account in terms of *primary contribution*, rather than employing some other similar "generative" relation(s). I hope this material will help persuade any initially skeptical readers that the primary contribution relation is sufficiently intelligible for the work I want it to do here—viz., illuminating the relation expressed by "is (un)lucky for."[23]

4. The Strokes Account of Lucky Events: Further Support and Defense

I turn now to providing further argument for, and defense of, the Strokes Account. Our extended versions of *Good Lottery* and *Bad Lottery* strongly confirm the account. Recall the extended version of *Good Lottery*, in which

---

[22] Cf. Sosa 2007, 95: "Something may explain the existence of a certain entity ... without even partially explaining why it has a given property. That it was made in a Volvo factory may explain the existence of a certain defective car, for example, without even partially explaining why it is now defective."

[23] Of course, "is a primary contributor to" (and so on) is vague. But that's not obviously a strike against analyzing luck in terms of primary contribution. Indeed, the observation may well count in favor of such an analysis, given the vagueness of "is (un)lucky for."

the lottery officials have just handed you the lottery's large cash prize. Here are two related facts:

- You were lucky that your birthday numbers won.
- You were lucky to receive the lottery prize.

The Strokes Account smoothly explains both facts. It was a stroke of good luck for you that your birthday numbers won, and you received the lottery prize primarily because your numbers won. Combined with these claims, the Strokes Account entails the above facts. That the Strokes Account so smoothly explains those facts counts significantly in its favor.

Now recall the extended version of *Bad Lottery*, in which the governor has just transferred your life savings into his checking account. Consider two related facts:

- You were unlucky that your birthday numbers "won."
- You were unlucky to lose your life savings to the governor.

Again, the Strokes Account easily explains both facts. It was a stroke of bad luck for you that your birthday numbers "won," and the governor took your savings primarily because your numbers "won." Combined with these claims, the Strokes Account entails the above facts. That the Strokes Account so easily explains those facts further confirms it's on the right track.[24]

My overall case for the Strokes Account will accumulate gradually throughout the remainder of the essay as we test the account against additional cases; enrich it with an analysis of *strokes of luck*; and show how the enriched account can properly handle all the earlier counterexamples to the leading theories of luck. I'll devote the rest of this section to defending the Strokes Account from some challenging objections to its left-to-right conditional. Let's start with this neat attempted counterexample:

> *Drawing Marbles*: An opaque jar holds ninety-nine red marbles and one green marble. Each trial consists of drawing, and then replacing, a single marble. You bet that the green marble will be drawn at least once over the course of 450 trials. The odds are squarely in your favor: there's about a 99 percent chance that at least one draw will produce the green marble. All 450 trials take place; alas, the green marble is never drawn.

---

[24] In light of cases like *Buried Treasure,* Steglich-Petersen (2010, 369) proposes this "ignorance" requirement on luck: E is at t lucky for S only if, just before t, S wasn't positioned to know E would occur then. Our lottery cases reveal that this proposal has counterintuitively skeptical consequences. The proposal implies, e.g., that just before I received the lottery prize, I wasn't in a position to know that I'd soon receive the prize.

You were unlucky that the green marble was never drawn. But this case doesn't seem to involve any strokes of bad luck for you. For, in each trial, there was only a 1 percent chance that the green marble would be drawn. True, in each trial, it would have been a stroke of good luck for you had the green marble been drawn. So in each trial, you failed to enjoy a stroke of good luck. But failing to enjoy a stroke of good luck doesn't suffice for suffering a stroke of bad luck. Consider: Winning the lottery is a stroke of good luck, but simply failing to win isn't itself a stroke of bad luck (cf. Levy 2011, 33). *Drawing Marbles* thus impugns the left-to-right conditional of the Strokes Account.[25]

Reply: The objector must be thinking that the complete 450-trial process couldn't *itself* be a stroke of bad luck for you. Why think that? The reasoning would presumably go something like this:

> None of the individual trials was itself a stroke of bad luck for you. And if none of the individual trials was a stroke of bad luck for you, then neither was the whole process made up of all those individual trials.

I accept the first premise but reject the second. The general thought underlying the second premise would have to be something like this:

> If E is a composite event none of whose individual parts was a stroke of good (bad) luck for S, then E isn't *itself* a stroke of good (bad) luck for S.

There are clear counterexamples to this principle. I am truly awful at darts. But I learn that I'll receive a big cash prize if I make a bull's-eye on my next throw. I try my hardest to make a bull's-eye; lo and behold, I succeed! We can understand this case so that making the bull's-eye was a stroke of good luck for me. Compatibly with such an understanding of the case, we can fill in the details so that each of the individual parts or "steps" of making the bull's-eye was itself highly likely to happen, given how things stood just before it happened. What emerges is a case in which making the bull's-eye was a stroke of good luck for me, even though none of the bull's-eye's individual parts or "steps" was *itself* a stroke of good luck for me.[26]

---

[25] Thanks to Georgi Gardiner, Maria Lasonen-Aarnio, and Doug MacLean for suggesting cases that inspired this one.

[26] *Objection: Drawing Marbles* is sufficiently structurally similar to *Buried Treasure* so that either each outcome (treasure find, 450 straight "red" draws) is lucky or neither is. But you've deemed one lucky and the other not. *Reply:* Actually, these cases are structurally quite different. *Drawing Marbles* features a "fragile" outcome of multiple individually "robust" phenomena. By contrast, *Buried Treasure* should be read as featuring a "robust" outcome of phenomena at least some of which were individually "fragile" (cf. Lackey 2008, 263–64). (Thanks to Lee Whittington for comments that led me to add this note.)

In the next section, I start developing an analysis of *strokes of luck* on which a composite event like making a bull's-eye or a multi-trial marble-drawing process can count as a stroke of good or bad luck for a subject. Once this analysis is on the table, I'll return to *Drawing Marbles* to verify that it implies that the whole 450-trial process is itself a stroke of bad luck for you. That will complete my defense of the Strokes Account from this case.[27]

Here's another, somewhat simpler attempted counterexample to the Strokes Account's left-to-right conditional. At the conclusion of the extended version of *Good Lottery*, you are lucky that you played your birthday numbers. But playing your birthday numbers needn't *itself* have been a stroke of good luck for you: we can safely suppose that there were no hidden obstacles to your selecting those numbers; that you played those numbers intentionally, knowingly, and freely; and so on. Further, we can suppose that your playing those numbers was not due primarily to any other stroke of good luck for you. So the Strokes Account's left-to-right conditional implies incorrectly that you are *not*, at day's end, lucky you played your birthday numbers.

Reply: Once we get clearer on what's true of this case, and how the Strokes Account bears on it, we'll see that the case doesn't threaten the account. Notice first that the critic expresses the objection's initial premise by way of a *present tense* ascription of luck regarding an *earlier* event: "... you *are* lucky that you *played* your birthday numbers." Now, like all the going accounts of luck set out above, the Strokes Account's analysandum is a relation that a subject S bears to an event E at a time t only if E happens (occurs, comes to obtain) *at t*. Initially, then, it's not at all clear that the Strokes Account's analysandum is the relation at play in the critic's first premise.

This lack of clarity raises a difficult question: How exactly we should interpret that first premise? This question is difficult because we can use *present tense* ascriptions of luck regarding *earlier* events to make any of a number of different claims. Notably, if given a "strict and literal" reading, any such ascription of luck is arguably false. For if an event has *already happened*, then there's a clear sense in which the event's occurrence is now "fixed"—this is the status labeled "the necessity of the past" or "accidental necessity" in the literature on human freedom and divine foreknowledge. But if an event's occurrence is now "fixed" in the relevant sense, then the event's occurrence is not now *lucky* for anyone—though, of course, it may well be that the event *was* lucky for someone *when it happened*.

[27] Thanks to Georgi Gardiner and Lee Whittington for helping me think through this objection.

A present tense ascription of luck regarding an earlier event might be either of the following (or something else—I needn't, and so won't try to, exhaust the possibilities here):

- A disguised *past tense* ascription of luck to the earlier event. For example: by uttering "I'm lucky I won the lottery," I can convey [Winning the lottery was lucky for me].
- A disguised *past tense* ascription of ignorance regarding some of the earlier event's later positive consequences (cf. Levy 2011, 20). For example: by uttering "I'm lucky I sold my stocks before the market crashed," I can convey [I unwittingly protected myself from the crash by selling my stocks].

Reflection on these two different uses of the relevant kind of luck ascription yields (what strike me as) the two most charitable interpretations of the critic's first premise. On the one hand, the critic's claim might be that *at the time you played your birthday numbers*, you were lucky to be playing those numbers. If so, then that premise is dubious: the very reasons given for thinking that playing your birthday numbers wasn't (when you played them) a stroke of good luck for you also support the claim that playing those numbers wasn't (when you played them) *lucky* for you. On the other hand, the critic might be claiming that, when you played your birthday numbers, you were unwittingly selecting what would turn out to be the lottery's winning numbers. If so, then while the pertinent premise is of course true, it's irrelevant to the Strokes Account. For on this second interpretation, the relation at play in that premise can't be the Strokes Account's analysandum: ignorance of a current act's later positive consequences is far too common a phenomenon to be the luck analyst's quarry. Upshot: Neither of the suggested interpretations of the critic's first premise yields a successful objection to the Strokes Account.

Having provided some initial support for and defense of the Strokes Account, I now move on to developing an analysis of *strokes of luck* that utilizes insights from the recent luck literature. As we'll see, when we plug this analysis into the Strokes Account, the result is a comprehensive new theory of luck that can handle all the earlier counterexamples to the literature's leading accounts.

5. Strokes of Luck: An Analysis and Some Important Implications

Bracketing a few bells and whistles that needn't detain us here, we get my favored analysis of *strokes of luck* by simply "copying-and-pasting" the right-hand-side of the Mixed Account:[28]

---

[28] For bells and whistles, see section 5 of Coffman 2007 and section 2.2 of Coffman forthcoming.

The Analysis: Event E is at t a stroke of good (bad) luck for subject S = df. (1) E is in some respect good (bad) for S, (2) E doesn't happen around t in a wide class of worlds close to the actual world before t, and (3) E isn't something S did intentionally.

A few initial remarks about the Analysis. First, the Analysis has it that whether E is at t a stroke of luck for S depends on how things stood just before t (i.e., all the way up to, but not including, t). That's intuitively plausible. Suppose that, while there was around the time of the Big Bang only a minuscule chance that E would eventually happen at t, there was just before t only a minuscule chance that E wouldn't happen at t. Suppose E happens at t. Given that there was, just beforehand, only a minuscule chance that E wouldn't happen then, E shouldn't count as a stroke of luck for anyone. Moral: Whether E is at t a stroke of luck for S depends (not on how likely it was at, for example, some point in the remote past that E would happen at t, but) on how likely it was just before t that E would happen then.[29]

Now let's take a closer look at condition (3). *Question:* Armed only with (3), can the Analysis capture all there is to the platitude that "strokes of luck are uncontrolled"? Or must we add more conditions to fully capture the "out of control" requirement on an event's being a stroke of luck for a subject? I'll explore this relatively large question by way of two comparatively smaller ones:

The Connections Question: Can the Analysis (as currently formulated) honor the most obvious connections between the concepts of *directly free action* and *stroke of luck*?[30]
The Abilities Question: How do strokes of luck relate to agents' abilities to produce/prevent events?[31]

Take the Connections Question first. It's extremely plausible to think that *directly* free acts—that is, free acts that don't owe their status as such to

---

[29] Following other theorists of luck (Pritchard [2005], Levy [2011]), here and elsewhere I employ a familiar notion of objective (metaphysical) chance on which the chance, just before t, that an event E will happen at t is equal to the "size" of the set of worlds close to actuality before t in which E happens at t, divided by the "size" of the whole set of worlds close to actuality before t (cf. van Inwagen [1997, 231ff.] and Williamson [2000, 123–24]).

[30] In section 2.1 of Coffman forthcoming, I consider how *stroke of luck* relates to *knowingly performed action.* Following Mele and Moser (1994, 45), I maintain that foreseen side-effects of intentional actions can count as things one does knowingly but not intention ally. If so, the Analysis leaves it open that something you did knowingly was also a stroke of luck for you. I describe a "side-effect action" case that illustrates this possibility.

[31] In line with my broad use of "do," I here use "produce" ("prevent") in a broad sense that covers both *causal production* (*causation of omission*) for events proper and *actualization* (*keeping from obtaining*) for states of affairs.

any other free acts (cf. Mele 2006, 87)—can't be strokes of luck for their agents.[32] Supposing that's right, can the Analysis honor this truth "as is"?

Yes, given the intuitively plausible assumption that any directly free act is something its agent does intentionally. Suppose that's right: an agent's act is directly free only if the agent does that act intentionally.[33] The left-to-right conditional of the Analysis entails that nothing an agent does intentionally is a stroke of luck for the agent. Therefore, given one intuitively plausible assumption about the nature of directly free action, the left-to-right conditional of the Analysis entails that nothing an agent does with direct freedom can be a stroke of luck for that agent.

On to the Abilities Question. Many theorists of luck endorse something along these lines: E is a stroke of luck for S only if S lacks "control over" E.[34] Unlike "control of," to say that you have "control over" E implies that you are free to produce and/or to prevent E (where S is "free to" produce/prevent E iff S both has it within her power and knows how to produce/prevent E). Many theorists thus endorse a requirement for E's being a stroke of luck for S on which S isn't both free to produce E and free to prevent E. Do strokes of luck really relate to agents' abilities as the theorists currently in focus seem to think?

Over the next few paragraphs, I'll argue for two theses:

Thesis 1: Possibly, E is a stroke of good luck for S, yet S is *both* free to produce E *and* free to prevent E.

Thesis 2: Possibly, E is a stroke of bad luck for S, yet S is *both* free to produce E *and* free to prevent E.

If Theses 1 and 2 are true, then E can be a stroke of (good or bad) luck for you despite the fact that you have "control over" E, in the sense of "control over" that proponents of the above popular construal of the "out of control" requirement on strokes of luck seem to have in mind. In light of the two upcoming examples that I'll call *Rigged Lottery* and *Two Buttons*, we should conclude that strokes of luck don't relate to agents' abilities as many theorists of luck seem to think.

---

[32] Something like the claim to which this note is appended animates the so-called Luck Argument(s) against libertarianism about metaphysical freedom (the thesis that metaphysical freedom exists and is incompatible with the truth of causal determinism). For a helpful critical survey of historical and contemporary "luck-driven" arguments against libertarianism, see Franklin 2011. I discuss such argumentation at length in chapters 5 and 6 of Coffman forthcoming.

[33] If a skeptical interlocutor pressed for further reasons to accept this claim, I'd lead with the following argument: Any directly free act is an act its agent does for a reason (cf. Levy 2011, 64). Any act an agent does for a reason is something its agent does intentionally (cf. Mele and Moser 1994, 64). Therefore, all directly free acts are done intentionally.

[34] Cf. Nagel 1976, Zimmerman 1987, Statman 1991, Greco 1995, Coffman 2007, 2009, and Levy 2009, 2011.

First, a case that supports Thesis 1:

> *Rigged Lottery*: Without your having had any say in the matter, you find
> yourself holding a ticket in a lottery that has been rigged in your favor.
> There are two ways you can win: press a button that will make you the
> (illegitimate) winner, or let the lottery proceed fairly in hopes that your
> number will be the one selected. For you to win either way, you *must*
> keep your ticket. Although you're free to make yourself the winner,
> you refrain from intervening. Indeed, given your character, there was
> almost no chance you'd exploit the set-up to win illegitimately. Lo and
> behold, you win fair and square!

Your *legitimate* lottery win was a stroke of good luck for you. But now
consider the fact that you won the lottery (period). You were free to make
yourself the winner, and you were also free to prevent your winning (by
destroying your ticket). So, if your winning (period) was a stroke of good
luck for you, then Thesis 1 is true. Was it indeed a stroke of good luck for
you that you won (period)?

If we try to argue for the conclusion that your winning was a stroke of
good luck for you, this reasoning will occur to us fairly early on:

> It was a stroke of good luck for you that you won *legitimately*. That you
> won legitimately entails that you won. The relation *being a stroke of
> good luck for* is "closed under entailment": if it was a stroke of good
> luck for you that P *and* P entails Q, then it was also a stroke of good
> luck for you that Q. Therefore, it was a stroke of good luck for you
> that you won.

But this argument fails, since *being a stroke of good luck for* clearly isn't
closed under entailment (cf. Dretske 1970, 1008–9). Suppose you legitimately win a large, normal lottery. Your legitimate win is a stroke of good
luck for you. That you legitimately win entails that there's a legitimate winner. But no one is lucky that there's a legitimate winner, not even you (that
there would be a legitimate winner was guaranteed).

Of course, the fact that the above argument fails is no reason to think
that your winning (period) in *Rigged Lottery* wasn't a stroke of good luck
for you. Indeed, although I myself have recently claimed that your winning
wasn't a stroke of good luck for you (cf. Coffman 2009, 503–4), I now feel
no inclination at all to say this, and instead feel a strong inclination to say
that your winning was in fact a stroke of good luck for you—as have many
people with whom I've discussed such cases.[35] So far as I can see at present,
then, *Rigged Lottery* establishes Thesis 1.

---

[35] Wayne Riggs has been especially influential on this front.

Turning to Thesis 2, consider this case:

> *Two Buttons*: You confront a device with two buttons, green and red. If you press the *red* button, you'll immediately suffer a painful sensation (100 percent chance). If you press the *green* button, there's a 99 percent chance you'll enjoy a pleasant sensation and only a 1 percent chance you'll suffer a painful sensation. You're free to do any of the following: press the red button; press the green button; omit pressing either button. Reasonably expecting to enjoy the pleasant sensation, you freely press the green button. You suffer a painful sensation.

Your suffering that painful sensation was a stroke of bad luck. Note that the Analysis entails as much: suffering the painful sensation was bad for you; in a wide class of worlds close to the actual world before you suffer the painful sensation, you don't suffer such a sensation; and, finally, suffering the painful sensation wasn't something you did intentionally. Nevertheless, you were free *both* to produce the painful sensation (you were free to press the red button) *and* to prevent it (you were free to omit pressing either button). So *Two Buttons* establishes Thesis 2.

We've been considering whether the Analysis can—armed only with condition (3)—capture what truth there is in the platitude that "strokes of luck are uncontrolled." We've made progress on that question by answering two comparatively smaller questions: the Connections Question and the Abilities Question. I've argued for an affirmative answer to the Connections Question: the Analysis can capture the most obvious connections between the concepts of *directly free action* and *stroke of luck*. And reflection on the Abilities Question has revealed that we shouldn't require, for an event's being a stroke of luck for you, that you lack *either* freedom to produce that event *or* freedom to prevent it. So reflection on the Abilities Question hasn't uncovered any new reason to think that the Analysis can't capture what truth there is in the pertinent platitude. On the basis of these answers to the indicated questions, then, I'm cautiously optimistic that condition (3) exhausts the "out of control" requirement on an event's being a stroke of luck for a subject.[36]

Call the conjunction of the Strokes Account with the Analysis the *Enriched Strokes Account* of lucky events. Now is as good a time as any to note that, since key notions involved in (1) and (2) of the Analysis admit of degrees, those conditions allow us to capture the fact that luck itself comes in degrees.[37] To illustrate: a proponent of the Enriched Strokes Account can say that some extremely positive (negative) and unlikely event was a huge

[36] Special thanks to Doug MacLean and Susan Wolf for extremely helpful conversation about the issues explored in this section.

[37] If acts can be done more or less intentionally—an issue I take no stand on here—then (3) also helps capture comparative facts about luck.

stroke of good (bad) luck for you, and that any positive (negative) event to which that huge stroke of good (bad) luck is a primary contributor is itself extremely (un)lucky for you; that some slightly positive (negative) and fairly likely event was only a small stroke of good (bad) luck for you, and that any positive (negative) event to which that small stroke of good (bad) luck is a primary contributor is only slightly (un)lucky for you; and so on. Next on the agenda is to verify that the Enriched Strokes Account vindicates some not-yet-fully-substantiated claims I relied on in earlier sections, and to defend the account from a challenging charge of incoherence.

6. The Enriched Strokes Account of Lucky Events: Further Support and Defense

In earlier sections, I relied on the following not-yet-fully-substantiated claims:

- In *Distracted Driver*, Evan's eventual safety was not itself a stroke of good luck for me (or for him). It was, however, a stroke of good luck for me (and for him) that I was *in a position* to save him, and lucky for me (and for him) that he ended up safe (section 2).
- In *Drawing Marbles*, it was a stroke of bad luck for you that the green marble wasn't drawn in any of the 450 trials (section 4).

Start with *Distracted Driver*. Just before saving Evan, I became *positioned* to save him. Once I became so positioned, it was *guaranteed* that I would soon save Evan. So Evan's eventual safety doesn't itself meet condition (2) of the Analysis. The Analysis thus implies that it wasn't a stroke of good luck for me (or for Evan) that he ended up safe. My becoming positioned to save Evan, however, was clearly a primary contributor to his eventual safety. Further, my becoming so positioned was good for me (and for Evan); we can safely suppose that it doesn't happen around the relevant time in a wide class of worlds close to the actual world beforehand; and it's not something that I (or Evan) did intentionally. The Enriched Strokes Account thus implies both that my becoming positioned to save Evan was a stroke of good luck for me (and for him) and that it was lucky for me (and for him) that he ended up safe.

Now recall *Drawing Marbles*. The fact that all 450 trials failed to produce a green marble was bad for you, and not something you did intentionally. So whether the trials' failure to produce a green marble was a stroke of bad luck for you comes down to whether there was—before the stretch of time over which all those individual trials took place (label it "t")—a sufficiently large objective chance that at least one trial would then produce a green marble. More precisely, we need to ask whether the following is true: In a wide class of worlds close to the actual world before t, at least one trial produces a green marble during t.

We should answer this question affirmatively. Recall that, before the trials took place, there was roughly a 99 percent chance that at least one trial would produce a green marble during t. So, in almost all of the worlds close to the actual world before t, at least *one* of the trials produces a green marble during t. So, not only was the trials' failure to produce a green marble bad for you, and not something you did intentionally; there was also, before the stretch of time over which it happened, a sufficiently large chance it *wouldn't* happen during that period. The Analysis thus vindicates the claim that the trials' failure to produce a green marble was *itself* a stroke of bad luck for you. More generally, the Analysis clearly and correctly allows for the possibility that some composite event is itself a stroke of (good or bad) luck for you, despite the fact that none of its individual parts or "steps" is.

I'll devote the rest of this section to defending the Enriched Strokes Account from a challenging objection inspired by both Latus (2003, 467ff.) and Levy (2011, 16–17, 29ff.). Suppose a subject, S, survives a game of Russian roulette in which only one of the revolver's six chambers is loaded. Alternatively, consider a subject S's being born with an exceptionally happy temperament. It seems that if the Analysis is correct, then we can understand S's case so that it doesn't involve *any* strokes of good luck for S. And if we *can* so understand S's case, yet S was nevertheless *lucky* to survive the game of roulette—or to be born with an exceptionally happy temperament—then the Strokes Account is false. But, the objection continues, S *was* lucky to survive the game—or to be born with an exceptionally happy temperament. Therefore, it seems that if the Analysis is true, then the Strokes Account is false. Upshot: The Enriched Strokes Account combines two *incompatible* views!

This objection rests on two basic premises:

Premise 1: If the Analysis is correct, then S's case needn't involve *any* strokes of good luck for S.

Premise 2: S was lucky to survive the roulette game—or, to be born with an exceptionally happy temperament.

I will argue that neither of the two envisaged cases is such that *both* premises are true of it.

Consider, first, S's surviving the roulette game. As it's described, there was roughly a 17 percent objective chance that S would not survive the game. That's a sufficiently large chance of death for S's survival to satisfy my intended reading of condition (2) of the Analysis.[38] So, contrary to the objector's suggestion, since S's survival is *also* in some respect good for S

---

[38] Contrast Coffman 2007, 390f., which I now deem mistaken.

and not something S did intentionally, the Analysis issues the intuitively correct verdict that S's survival was indeed a stroke of good luck for him.

Now, the objector might try to sidestep this reply by invoking what I'll call the

> *Inverse Proportionality Thesis (IPT)*: The degree of chanciness required
> for an event to count as lucky for one is *inversely proportional* to the
> degree to which the event is good for one.

Latus (2003, 467–68) explicitly endorses IPT.[39] If IPT is true, then we can simply modify the Russian roulette case so that there was only a minuscule chance that S would die. So modified, S's survival will be too probable to satisfy condition (2) of the Analysis, yet (claims the critic) will still count as lucky.

Far from revitalizing the above argument, such a move would actually render it irrelevant to the Strokes Account. To begin to see this, note that luck *runs riot* under IPT: for any event E that's extremely significant for—but not done intentionally by—S, E will count as *lucky* for S provided that there was at least an extremely small objective chance that E wouldn't happen. There's an endless supply of examples that illustrate how luck proliferates given IPT; we'll consider just one, picked more or less at random. I recently survived a routine flight from Chicago to South Bend. Surviving that flight was extremely significant for me, and didn't count as something I did intentionally. Further, we can suppose that there was at least an extremely small objective chance I wouldn't survive the flight (think: mechanical problems, pilot errors, wayward geese, blood clots, Large Hadron Collider catastrophe, …). IPT implies that I was lucky to survive my routine flight from Chicago to South Bend. More generally, IPT entails that a truly staggering number and variety of events are lucky for us.

But this means that, whatever relation is at play in IPT (assuming that the principle is indeed true), it's *not* the Strokes Account's analysandum (cf. Lackey 2008, 258). The luck analyst's quarry, remember, is a phenomenon one paradigm of which is your winning a large fair lottery, and the presence of which can arguably keep true beliefs from constituting knowledge, rational actions from qualifying as free, and so on. Accordingly, the luck analyst's quarry simply can't be as pervasive as is the relation at play in IPT. Tying the above argument to IPT would therefore render it irrelevant to the Enriched Strokes Account. I conclude, then, that however exactly we fill in

---

[39] According to Levy (2011, 17), "the degree of chanciness necessary for an event to count as lucky is *sensitive* to the significance of that event for the agent" (my emphasis). However, Levy (2011, 17) also explicitly agrees with me that "there is a threshold, even for an event as significant for the agent as the death [S] risks, below which surviving is merely fortunate and not lucky (suppose that the probability of the gun's firing was 0.00001%)."

its details, the Russian roulette case isn't such that both Premises 1 *and 2* are true of it.

On to the "happy temperament" case. From such claims of "constitutive luck" as that S was lucky to be endowed with an exceptionally happy temperament, Levy (2011, 32ff.) infers that the following condition suffices for your being lucky relative to some positive character trait:

> Your having the pertinent trait is good for you, you lack control over having the trait, and the trait is sufficiently rare across human experience.

I agree with Levy that you should think some such condition suffices for luck with respect to some positive trait *if* you deem S lucky to have his exceptionally happy temperament. But we now have all we need, I believe, for a reductio of the claim that S is lucky to have such a temperament (cf. Rescher 1995, 28–29). Suppose that, on the basis of firmly held intentions formed long ago, S's parents intentionally "engineered" S so as to *ensure* that S would be "hard-wired" with an exceptionally happy temperament. We can easily fill in the details so that S isn't *lucky* to have his exceptionally happy temperament.[40] But just as clearly, S's having that rare positive trait may be good for, and uncontrolled by, S. So the proposed sufficient condition for luck relative to some positive trait turns out on closer inspection to be *in*sufficient. Finally, since we were led to that mistaken condition by the claim that S is—in the original, "manipulation-free" case—lucky to have an exceptionally happy temperament, we should conclude that S is *not* in fact lucky to have such a temperament in the original case.

In response, Levy might concede that in the "manipulation-involving" case, S isn't lucky to have the relevant sort of temperament, but then weaken the above alleged sufficient condition by replacing its antecedent with the following, logically stronger proposition:

> Your having the pertinent trait is good for you, *no one* has direct control over your having the trait, and the trait is sufficiently rare across human experience.[41]

The revised sufficient condition will be immune to "manipulation-involving" examples. Unfortunately, though, there are plausible counterexamples even to the revised condition. Suppose it's at least (conceptually) *possible* that there be a divine human being.[42] If there *were* such a being, she would *essentially* or *necessarily* exemplify such features as omniscience,

---

[40] Cf. Levy's (2011, 21–22) own claims about the case he calls "Buried Treasure*," discussed in section 2 above.

[41] Thanks to Lee Whittington for suggesting that such a move should be discussed.

[42] Morris 1986 is a "contemporary classic" in defense of this possibility.

omnipotence, and omnibenevolence. Such traits would be good for their possessor, and rare across human experience. So, provided that no one has control over properties a thing exemplifies *essentially* or *necessarily*, the weaker putative sufficient condition implies that a divine human being would be *lucky* to be omniscient, omnipotent, and omnibenevolent. But that's a mistake: if there *were* a divine human being, she wouldn't be *lucky* to have such traits.[43] Therefore, the weaker alleged sufficient condition for luck relative to positive traits isn't credible enough to ground a successful objection to the Enriched Strokes Account.

The upshot of the last few paragraphs is that, while Premise 1 may be true of the "happy temperament" case, Premise 2 isn't. And yet there does seem to be a kernel of truth in many such "constitutive luck" claims. Let me offer a pair of complementary error theories for such claims.

To begin to see the first error theory, consider the important distinction between *luck* and *fortune*. Whereas an event is *(un) lucky* only if it's properly related to a stroke of luck, you can be *(un)fortunate* relative to an event even if there are no strokes of luck in its vicinity. Rescher helpfully draws the luck/fortune distinction in these passages:

> You are fortunate if something good happens to or for you in the natural course of things. But you are lucky when such a benefit comes to you despite its being chancy. … Fate and fortune relate to the conditions and circumstances of our lives generally, luck to the specifically chancy goods and evils that befall us. Our innate skills and talents are matters of good fortune; the opportunities that chance brings our way to help us develop them are for the most part matters of luck… .
>
> The positive and negative things that come one's way in the world's ordinary course—including one's heritage (biological, medical, social, economic), one's abilities and talents, the circumstances of one's place and time (be they peaceful or chaotic, for example)—all these are matters of what might be characterized as fate and fortune. People are not unlucky to be born timid or ill-tempered, just unfortunate. But the positivities and negativities that come one's way by chance and unforeseen happenstance … are matters of luck. (Rescher 1995, 28–29)

Despite the fact that luck and fortune can be so distinguished on reflection, they are also (and obviously) closely related—for example, like (un)lucky events, (un)fortunate events must be good (bad) for you, and also in some sense "uncontrolled." Given the close similarity between luck and fortune, we shouldn't be too surprised to find theorists sometimes running them together. With the luck/fortune distinction in hand, then, we can plausibly conjecture that Levy has confused these notions in claiming that S is *lucky* to have an exceptionally happy temperament. What's true is that S is (merely) *fortunate* to have his exceptionally happy temperament—S has been "blessed with" such a temperament.

---

[43] Notably, Levy (2011, 30) concedes the plausibility of the claim that one isn't lucky relative to properties one exemplifies *necessarily* or *essentially.*

A second error theory for Levy's judgment about the temperament case invokes the familiar distinction between (a) an assertive utterance's having true propositional content and (b) the utterance's conveying some or other true proposition(s) similar to its actual propositional content. We can imagine circumstances in which assertively uttering a sentence of the form "S is lucky to have his exceptionally happy temperament" conveys one or another of a range of similar (to the utterance's actual content) truths more effectively than would a counterpart utterance that replaces "lucky" with "fortunate." Such conveyed truths might include any of the following:

- *In anyone's mouth*: S hasn't done anything to earn (deserve, merit) such a temperament.
- *In S's mouth*: S is grateful to have such a temperament; S recognizes that many people fail to enjoy such a temperament through no fault of their own.
- *In someone else's mouth*: S should be grateful to have such a temperament; S should recognize that many people fail to enjoy such a temperament through no fault of their own.

Since assertive utterances of sentences of the form "S is lucky to have his exceptionally happy temperament" can very effectively convey any of the above (and, in all likelihood, various other) kinds of truths, it's quite understandable how one could end up thinking that the actual propositional content of such an utterance is true—even if, as I've suggested, the proposition such an utterance strictly expresses is false. Perhaps Levy's (and others') sense that such an utterance's actual propositional content is true stems from conflating that content with some or other similar truth(s) conveyed or implicated—though not strictly expressed—by the utterance.[44]

This concludes my defense of the Enriched Strokes Account from the challenging objection discussed over the past several paragraphs. In the next section, I draw things to a close by revisiting our earlier counterexamples to the literature's leading theories of luck to show that the Enriched Strokes Account can properly handle all of them.

7. The Enriched Strokes Account and the Counterexamples to the Leading Theories

There are six cases to discuss here; I'll take them in order of appearance. First, *Buried Treasure*, which shows that condition (ii) of the Modal and Mixed Accounts isn't required for an event's being lucky for an agent.

---

[44] Similar points can be applied to paradigm ascriptions of "circumstantial luck" (cf. Nagel 1976)—for example, "lucky to have been outside Germany when the Nazis came to power," "lucky to have such a wonderful family," and so on. For a complementary perspective on such ascriptions of luck that also takes into account fascinating recent psychological work on luck ascriptions, see Pritchard and Smith 2004, especially pp. 12 and 24.

The Enriched Strokes Account vindicates the intuitive claim that Vincent's lucky discovery was due primarily to a stroke of good luck for him—viz., his forming an intention whose execution was counterfactually sufficient for discovering the treasure (that is, an intention such that he would discover the treasure were he to execute it). The Analysis implies that Vincent's forming such an intention was a stroke of good luck for him: forming such an intention was good for Vincent; it wasn't something he did intentionally;[45] and—we can safely suppose—it doesn't happen around the relevant time in a wide class of worlds close to the actual world just beforehand (more on this claim in a moment). Further, since Vincent's forming such an intention is clearly a primary contributor to his subsequent discovery (which, obviously, is in some respect good for him), the Strokes Account implies that his discovery is itself lucky. The Enriched Strokes Account can deliver the correct verdict on *Buried Treasure*.

Pointed question: How is it "safe" to suppose that, in a wide class of worlds close to the actual world before Vincent makes the relevant choice, he doesn't so choose around then? Here, it's essential to keep in mind that Lackey (2008, 261–62) offers *Buried Treasure* as a clear counterexample to the alleged necessity of the Modal (and Mixed) Account's condition (ii) for an event's being lucky, and that I'm accepting the case as such. Obviously, if *Buried Treasure* is to have any chance of achieving Lackey's aim, it must elicit from us a sufficiently strong sense that Vincent's discovery is lucky. Charity thus demands that we (try to) interpret *Buried Treasure* so that it elicits such a sense from us. Now, consider a reading of Lackey's example in which antecedent factors—personal and/or impersonal—align so as to make Vincent's *choice* (as opposed to the later *discovery*) appear around the time he actually makes it in the vast majority of worlds close to the actual world just beforehand. I submit that, once we're forced to so interpret the case, whatever sense we may initially have had that Vincent's treasure find was *lucky* will start to evaporate—the discovery will instead start seeming "fated." I read the following portion of Lackey's commentary on *Buried Treasure* as an attempt to forestall precisely the sort of interpretation now under consideration: "Counterfactual robustness [of Vincent's discovery] is ensured in BURIED TREASURE through absolutely no deliberate intervention of any sort; instead, circumstances *just happen to fortuitously combine* in such a way so as to make Vincent's discovery appear both in the actual world and in all of the relevant nearby worlds. Indeed, it is precisely because of this fortuitous combination of circumstances that the discovery of the buried treasure is so clearly a lucky event" (Lackey 2008, 263). We should assume, then, that there are no antecedent factors aligning so

---

[45] Note that I'm *not* claiming that Vincent didn't intentionally choose to plant roses in his mother's memory on the northwest part of the island. Rather, my claim here is just that Vincent didn't intentionally form an intention possessed of the relevant counterfactual property—*viz., being such that he'd find buried treasure were he to execute it.*

as to make Vincent's *choice* appear around the time he actually makes it in the vast majority of worlds close to the actual world just beforehand. And this means that we can indeed safely suppose that Vincent's choice doesn't happen around the relevant time in at least a wide class of relevant worlds.

Next, the Oxfam donation case, which refutes the Modal Account's right-to-left conditional. Since choosing to make the donation is something you did intentionally, the Analysis entails that your choice to donate wasn't itself a stroke of good luck for you. Moreover, the case is naturally understood so that your choice wasn't due primarily to some other stroke of good luck for you. Unlike the Modal Account, then, the Enriched Strokes Account delivers the correct verdict that your choosing to donate was not lucky for you.

On to *African Expedition*, which also shows that (ii) of the Modal and Mixed Accounts isn't required for an event's being lucky for an agent. The Enriched Strokes Account honors Riggs's suggestion (2009, 217) that visiting the pertinent area during a period when there would be a life-saving eclipse was lucky for *Smith*, but not for *Jones*. Smith visited that area during a period when there would be a life-saving eclipse primarily because he'd earlier become constrained to visit the area during the indicated period. The Analysis implies that Smith's becoming constrained to visit the area during that period was a stroke of good luck for him: his becoming so constrained was clearly good for him; it wasn't something he did intentionally; and, we can safely suppose, it doesn't happen around the relevant time in a wide class of worlds close to the actual world just beforehand.[46] Therefore, since Smith's visiting the area during a period when there would be a life-saving eclipse was clearly good for him, the Enriched Strokes Account implies correctly that Smith's visiting the area during a period when there would be a life-saving eclipse was lucky for him. Note, finally, that the Enriched Strokes Account also implies correctly that *Jones* wasn't lucky to visit the area during such a period, for the case harbors no strokes of good luck for *him*.[47]

We finish with three relatively "easy" examples: Katelyn and her solar-powered underground facility, *Kidnapping*, and *Distracted Driver*. The first of these examples refutes Riggs's Control Account's right-to-left conditional. The Enriched Strokes Account, by contrast, correctly implies that the sun's rising on the morning in question wasn't lucky for Katelyn. The

---

[46] In line with the recent discussion of *Buried Treasure,* it's worth noting that if we interpret *African Expedition* in such a way that antecedent factors aligned so as to *guarantee* that Smith would become so constrained around the relevant time, then whatever sense we may initially have had that Smith was *lucky* to visit the area during a period when there would be a life-saving eclipse will start evaporating—his visiting the area then will start seeming fated.

[47] Wasn't Smith's becoming so constrained also a stroke of luck for *Jones*? No, for Smith's *becoming* so constrained wasn't in any way good for Jones (at that point, recall, Smith hadn't yet invited Jones on the trip). True, Smith's *being* so constrained when issuing his invitation was good for Jones. But since there was just beforehand no chance that Smith wouldn't be so constrained then, Smith's being so constrained was not a stroke of good luck for anyone.

Analysis entails that the sunrise wasn't itself a stroke of luck for Katelyn: it was too objectively likely to meet condition (2). And the case is naturally understood so that the sunrise wasn't due primarily to some other stroke of luck for Katelyn. Unlike the Control Account, then, the Enriched Strokes Account issues the correct verdict that the sun's rising on the morning in question was not lucky for Katelyn.

Kidnapping, recall, shows that an event can be lucky for you *even if* you successfully exploit the event for some purpose—and so, that the Control Account's condition (ii) isn't required for an event's being lucky for a subject. The Enriched Strokes Account, by contrast, entails that my lottery win in this case was a stroke of good luck for me, despite the fact that I exploited it for some purpose (viz., to get Zach back): the win was good for me; it wasn't something I did intentionally; and it doesn't happen around the relevant time in a wide class of worlds close to the actual world just beforehand. Finally, as for *Distracted Driver*—which shows that condition (iii) of the Control and Mixed Accounts isn't necessary for an event's being lucky for an agent—I've already argued that the Enriched Strokes Account issues the correct verdicts on this case (section 6): despite being something I brought about intentionally, Evan's eventual safety was lucky for me; for his eventual safety was due primarily to my becoming positioned to save him, which was itself a stroke of good luck for me.

This completes my argument that the Enriched Strokes Account can properly handle all the cases that make trouble for the literature's leading theories of luck. We therefore have strong reason to think that (what I earlier dubbed) Lackey's Inference is indeed hasty. We can (and should) agree with Lackey that cases like *Buried Treasure* and *Distracted Driver* show that conditions (ii) and (iii) of the Mixed Account are unnecessary for an event's being lucky for a subject. But I've now presented a strong cumulative case for the Enriched Strokes Account, which—while consistent with Lackey-type counterexamples to the leading theories of lucky events—entails that conditions (ii) and (iii) are nevertheless (in Lackey's words) "distinctive of, and central to, the concept of luck." For, according to the Enriched Strokes Account, (ii) and (iii) are requirements on an event's being a *stroke of luck*, a notion on which the concept *lucky event* is parasitic.

Acknowledgments

Special thanks to Nena Davis, Georgi Gardiner, Trevor Hedberg, Albert Hu, Neil Levy, David Palmer, Carolina Sartorio, and Lee Whittington for helpful written comments on earlier drafts. Thanks also to Devon Bryson, Randy Clarke, Tomis Kapitan, Maria Lasonen-Aarnio, John Lemos, Doug MacLean, Garrett Pendergraft, Duncan Pritchard, Wayne Riggs, Kate Ritchie, Tom Senor, Susan Wolf, and members of the University of Tennessee Philosophy Department Research Seminar for helpful discussion of various issues explored here. Finally, I'm very grateful to the University of Tennessee for generously supporting my work on this essay through both a Humanities Center Faculty Fellowship and a Faculty Development Leave.

References

Ballantyne, Nathan. 2012. "Luck and Interests." *Synthese* 185:319–34.

Bennett, Jonathan. 1988. *Events and Their Names*. Indianapolis: Hackett.

Coffman, E. J. 2007. "Thinking About Luck." *Synthese* 158:385–98.

———. 2009. "Does Luck Exclude Control?" *Australasian Journal of Philosophy* 87:499–504.

———. Forthcoming. *Luck: Its Nature and Significance for Human Knowledge and Agency*. New York: Palgrave Macmillan.

DeRose, Keith. 2009. *The Case for Contextualism*. Oxford: Oxford University Press.

Dretske, Fred. 1970. "Epistemic Operators." *Journal of Philosophy* 67: 1007–23.

Franklin, Christopher. 2011. "Farewell to the Luck (and Mind) Argument." *Philosophical Studies* 156:199–230.

Greco, John. 1995. "A Second Paradox Concerning Responsibility and Luck." *Metaphilosophy* 26:81–96.

Kim, Jaegwon. 1974. "Noncausal Connections." *Noûs* 8:41–52.

Lackey, Jennifer. 2008. "What Luck Is Not." *Australasian Journal of Philosophy* 86:255–67.

Latus, Andrew. 2003. "Constitutive Luck." *Metaphilosophy* 34:460–75.

Levy, Neil. 2009. "What, and Where, Luck Is: A Response to Jennifer Lackey." *Australasian Journal of Philosophy* 87:489–97.

———. 2011. *Hard Luck: How Luck Undermines Free Will and Moral Responsibility*. Oxford: Oxford University Press.

Lewis, David. 1986. "Causation." In *Philosophical Papers II*. Oxford: Oxford University Press.

Mele, Alfred. 2006. *Free Will and Luck*. Oxford: Oxford University Press.

Mele, Alfred, and Paul Moser. 1994. "Intentional Action." *Noûs* 28:39–68.

Morris, Thomas. 1986. *The Logic of God Incarnate*. Ithaca, N.Y.: Cornell University Press.

Nagel, Thomas. 1976. "Moral Luck." *Proceedings of the Aristotelian Society* 76:136–50.

Paul, L. A., and Ned Hall. 2013. *Causation: A User's Guide*. Oxford: Oxford University Press.

Pritchard, Duncan. 2005. *Epistemic Luck*. Oxford: Oxford University Press.

Pritchard, Duncan, and Matthew Smith. 2004. "The Psychology and Philosophy of Luck." *New Ideas in Psychology* 22:1–28.

Rescher, Nicholas. 1995. *Luck: The Brilliant Randomness of Everyday Life*. New York: Farrar, Straus and Giroux.

Riggs, Wayne. 2007. "Why Epistemologists Are So Down on Their Luck." *Synthese* 158:329–44.

———. 2009. "Luck, Knowledge, and Control." In *Epistemic Value*, ed. Adrian Haddock, Alan Millar, and Duncan Pritchard, 204–21. Oxford: Oxford University Press.

Sosa, Ernest. 2007. *A Virtue Epistemology*. Oxford: Oxford University Press.

Statman, Daniel. 1991. "Moral and Epistemic Luck." *Ratio* 4:146–56.

Steglich-Petersen, Asbjorn. 2010. "Luck as an Epistemic Notion." *Synthese* 176:361–77.

van Inwagen, Peter. 1983. *An Essay on Free Will*. Oxford: Oxford University Press.

———. 1997. "Against Middle Knowledge." *Midwest Studies in Philosophy* 21:225–36.

Williamson, Timothy. 2000. *Knowledge and Its Limits*. Oxford: Oxford University Press.

Zimmerman, Michael. 1987. "Luck and Moral Responsibility." *Ethics* 97:374–86.

CHAPTER 3

LUCK ATTRIBUTIONS AND COGNITIVE BIAS

STEVEN D. HALES AND JENNIFER ADRIENNE JOHNSON

1. Introduction

Luck plays an important role in several philosophical debates. In epistemology there is the issue of epistemic luck—how the presence of luck undermines the connection a belief has to the truth and thereby prevents the belief from becoming knowledge. In ethics there is the problem of moral luck, the matter of how to morally evaluate two agents who are in the same situation with the same intentions, but one agent's actions lead to bad consequences and the other agent's actions do not, when the sole difference is that the first was subject to bad luck. Political philosophers worry about luck egalitarianism, the view that injustice is to be partly understood as variations in luck in the social and genetic lotteries. There is also the luck problem in free will. Libertarians about free will hold that if an agent performs an action A, she might have performed action B instead, even given the same past and the laws of nature. The puzzle is that then the performance of A seems to be a matter of luck, since there was nothing that determined it. In the philosophy of science there is the matter of how to understand serendipity, and the role that it plays in the logic of scientific discovery.

Philosophers have developed three different theories of luck to help address these various concerns. The first is the probability theory, according to which an occurrence is lucky (or unlucky) only if it was improbable to occur (Bewersdorff 2005; Ambegaokar 1996; Rescher 1995). The second theory of luck is the modal theory, according to which an event is lucky only if it is fragile—had the world been very slightly different it would not have occurred (Pritchard 2005 and 2014; Levy 2011; Teigen 2005). The third

The Philosophy of Luck, First Edition. Edited by Duncan Pritchard and Lee John Whittington.
Chapters © 2014 Metaphilosophy LLC and John Wiley & Sons Ltd, except for "Luck as Risk and the Lack of Control Account of Luck" © 2015 Metaphilosophy LLC and John Wiley & Sons Ltd. Book compilation © 2015 Metaphilosophy LLC and John Wiley & Sons Ltd.
Published 2015 by John Wiley & Sons Ltd.

theory of luck is the control view, which states that if a fact was lucky or unlucky for a person, then that person had no control over whether it was a fact (Mele 2006; Levy 2011; Greco 2010).

In this essay we argue that all three theories of luck face serious challenges from experimental psychology. In our own experimental work discussed below, we show that the luck attributions of naïve participants are shot through with various cognitive biases. We then argue that philosophical theories of luck cannot adequately accommodate these empirical results. If this is correct, then the existence of pervasive bias raises the possibility that there is no such thing as luck. It may be that attributions of luck are a form of post hoc storytelling, or even mythmaking; that they are merely a narrative device used to frame stories of success or failure. Perhaps luck is analogous to pareidolia, our innate tendency to find visual patterns in random data, and events are lucky to the same extent that automobiles have faces, or a grilled cheese sandwich looks like the Virgin Mary. A rejection of luck as a genuine fact of the world would have far-reaching consequences in various philosophical domains.

We developed three primary hypotheses. Our first hypothesis was that the luck attributions of naïve participants would be subject to the cognitive bias of framing. In this case, we are operationally defining "framing" as a change in the wording of a problem. The literature clearly indicates that changing the framing of a problem influences decision making. For example, Tversky and Kahneman (1981, 453) presented the following scenario to participants:

> Imagine that the U.S. is preparing for the outbreak of an unusual Asian disease, which is expected to kill 600 people. Two alternative programs to combat the disease have been proposed. Assume that the exact scientific estimate of the consequences of the programs is as follows:
>
> If Program A is adopted, 200 people will be saved.
>
> If Program B is adopted, there is $^1/_3$ probability that 600 people will be saved, and $^2/_3$ probability that no people will be saved.
>
> Which of the two programs would you favor?

Even though the outcomes of Program A and B are statistically equivalent, 72 percent of participants chose a guaranteed gain (Program A) compared to the risk of saving none (Program B, chosen by only 28 percent of participants). In this case, and many other examples like it, people often fail to notice that the two outcomes have identical deep structures, and they are influenced instead by the surface features of the scenario. Likewise, we expected the same to be true when participants judged the luckiness of scenarios with a single deep structure but different surface features (see table 1 for scenarios). That is, positive framing of the scenario (for example, Tara **hit five** out of six numbers in the lottery) would be considered more lucky

TABLE 1. Luck framing in short vignettes. Words in **bold** signify the lucky versions, and *italic words in parentheses* signify the unlucky versions

Vignette 1	Tara Cooper **hit five** (*missed one*) out of six numbers in the Megabuck$ lottery.
Vignette 2	Mark Zabadi, new to the game of basketball, shot ten free throws and **made five** (*missed five*) of them.
Vignette 3	A severe snowstorm hit the town. Half of the town's residents **never lost** (*lost*) their power.
Vignette 4	Vicki Mangano, a casual bowler, almost bowled a perfect three hundred game. She **hit eleven strikes in a row** (*missed two pins in the last frame*) to end with a 298.
Vignette 5	Derek Washington **walked away without a scratch** (*was nearly killed*) when his car was destroyed by a 150-pound tractor-trailer tire.
Vignette 6	A tornado swept through an Oklahoma town, leaving many buildings in shambles. Half the shops on main street were **spared** (*destroyed*).
Vignette 7	James Goldberg's car slid on an icy road and **just missed hitting** (*nearly hit*) a pedestrian.
Vignette 8	In the first baseball game of the season, José Ramirez had four chances at bat and got **on base** (*out*) twice.

than negative framing of the same exact outcome (e.g., Tara **missed one** out of six numbers in the lottery).

Framing of a problem can also involve altering the background information or context of a scenario. Again, seminal work by Kahneman and Tversky (1984) showed how problems with the same deep structure will be judged differently based on the framing of their contexts. Consider these two problems from Kahneman and Tversky (1984, 347):

Problem 1: Imagine that you have decided to see a play and paid the admission price of $10 per ticket. As you enter the theater, you discover that you have lost the ticket. The seat was not marked, and the ticket cannot be recovered. Would you pay $10 for another ticket?

Problem 2: Imagine that you have decided to see a play where admission is $10 per ticket. As you enter the theater, you discover that you have lost a $10 bill. Would you still pay $10 for a ticket for the play?

In both problems, there is a loss of $10. Yet, only 46 percent of participants would purchase a new ticket in Problem 1, while 88 percent would purchase a new ticket in Problem 2. The framing of the context influenced participants' decision making. Likewise, we expected that changing the context of the luck scenarios would change perceptions of luckiness. We created short, one-sentence vignettes but also created long, four- or five-sentence vignettes with more context built in to them (see table 2). Therefore, our second hypothesis was that the longer vignettes, with more background information regarding the scenario, would show a greater framing effect

TABLE 2. Luck framing in long vignettes. Words in **bold** signify the significant critical event. All other words are identical between positively and negatively framed vignettes

	Positively framed long vignette	Negatively framed long vignette
Vignette 1	**"I hit five out of six! I've never come anywhere close to hitting the big jackpot before!** It was just unbelievable," Cooper exclaimed, still stunned. Berwick bakery worker Tara Cooper stopped off at her usual place for a breakfast coffee and bagel, Brewed Awakening, and decided to pick up a lottery ticket before heading to first shift. "I don't usually play Megabuck$, and don't know why I did today." After work, she checked her numbers online. "I was like, oh my God!"	**"I missed the jackpot by one lousy number! Story of my life.** It was just unbelievable," Cooper exclaimed, still stunned. Berwick bakery worker Tara Cooper stopped off at her usual place for a breakfast coffee and bagel, Brewed Awakening, and decided to pick up a lottery ticket before heading to first shift. "I don't usually play Megabuck$, and don't know why I did today." After work, she checked her numbers online. "I was like, oh my God."
Vignette 2	**"I hit half my shots from the free throw line! Not bad for a beginner, huh?" Mark exclaimed with a grin.** Even though he was one of the tallest kids in his class, Mark Zabadi had never picked up a basketball before. "I dunno," he said, "Guess I'm more of a gamer—not much of a team sports guy." But when some of his friends found themselves short a player for a pickup game, they convinced Mark to play.	**"Yeah, I missed half my shots from the free throw line. Not great, huh?" Mark said with a frown.** Even though he was one of the tallest kids in his class, Mark Zabadi had never picked up a basketball before. But when some of his friends found themselves short a player for a pickup game, they convinced Mark to play. "I dunno," he said, "Guess I'm more of a gamer—not much of a team sports guy."
Vignette 3	**"Half of the residents never lost their power,"** reported the mayor. **"It could have been a lot worse. We dodged a bullet."** Roads were slick for morning commuters, and icy trees knocked out electrical lines after a major winter storm blanketed the area in snow and ice this past weekend. Forecasters had predicted that the town would take the brunt of the worst storm of the season.	**"Half of the residents lost their power,"** reported the mayor. **"It can't get much worse. We weren't able to dodge this bullet."** Roads were slick for morning commuters, and icy trees knocked out electrical lines after a major winter storm blanketed the area in snow and ice this past weekend. Forecasters had predicted that the town would take the brunt of the worst storm of the season.
Vignette 4	**Last night Vicki Mangano bowled a 298 by hitting eleven strikes in a row, by far her best game ever.** Her teammates, The Rolling Rocks, were taking her out for pizza and beer afterward to celebrate. "I just couldn't miss!" she exclaimed. "I was totally in the zone." One of Vicki's teammates joked, "I just wish some of that lightning would strike me too."	Last night, Vicki Mangano **just missed out on bowling a perfect game after missing two pins** in the last frame. Her teammates, The Rolling Rocks, were taking her out for pizza and beer afterward to try to cheer her up. "I just couldn't miss," she said disappointedly. "I was totally in the zone." Said one of Vicki's teammates, "I just hope some of that lightning doesn't strike me too."

TABLE 2. (*Continued*)

	Positively framed long vignette	Negatively framed long vignette
Vignette 5	**"I walked away without a scratch! I must have a guardian angel. All my friends told me I should buy a lottery ticket tonight,"** said accident survivor Derek Washington. Washington was driving down the interstate when a loose tractor-trailer tire barreled into oncoming traffic. The massive 150-pound tire peeled back the roof of his Camry like a tin can and shattered his windshield.	**"I was nearly killed! I'm driving to work, minding my own business, and my car is totally destroyed in some freak accident,"** said victim Derek Washington. Washington was driving down the interstate when a loose tractor-trailer tire barreled into oncoming traffic. The massive 150-pound tire peeled back the roof of his Camry like a tin can and shattered his windshield.
Vignette 6	**"Half of my buildings look like nothing happened at all. They survived without losing a shingle,"** said Michelle Simmons. "There's no rhyme or reason to it. It's weird how the tornado just seemed to dance around." Simmons is a fifth-generation Oklahoma resident who owns several commercial rental properties right in the bull's-eye of the tornado's 220-mph winds. The governor of Oklahoma declared a state of emergency for the central part of the state after yesterday's F4 twister.	**"Half of my buildings have been completely flattened. There's nothing but some shattered framing and pipes left,"** said Michelle Simmons. "There's no rhyme or reason to it. It's weird how the tornado just seemed to dance around." Simmons is a fifth-generation Oklahoma resident who owns several commercial rental properties right in the bull's-eye of the tornado's 220-mph winds. The governor of Oklahoma declared a state of emergency for the central part of the state after yesterday's F4 twister.
Vignette 7	**"I fishtailed and just missed hitting into this guy walking to the pizza place. He moved out of the way just in time,"** said local driver James Goldberg. Drivers are well advised to look out for black ice, especially on the minor roads. The light drizzle on top of below-freezing ground temperatures has made the roads as slick as Teflon. The weather is expected to improve tomorrow.	**"I hit a patch of black ice and practically killed this guy who was walking to the pizza place. One minute the road is fine and the next I almost run over a guy,"** said local driver James Goldberg. Drivers are well advised to look out for black ice, especially on the minor roads. The light drizzle on top of below-freezing ground temperatures has made the roads as slick as Teflon. The weather is expected to improve tomorrow.
Vignette 8	**"I'm feeling good about this season,"** said center fielder José Ramirez, who **got on base twice in four at-bats** yesterday in the season opener against Cleveland. "First game of the year. I trained so hard in the off-season. For whatever reason the pitches all looked slow to me today. It was like playing t-ball again. I can't believe it!"	**"I should have done better,"** said center fielder José Ramirez, who got out twice in four at-bats yesterday in the season opener against Cleveland. "First game of the year. I trained so hard in the off-season. For whatever reason the pitches all looked slow to me today. It was like playing t-ball again. I can't believe it."

TABLE 3.   Examples of positively framed long vignette with critical event moved from beginning to middle to end

Vignette 1 (long version) Positive event at beginning	Vignette 1 (long version) Positive event in middle	Vignette 1 (long version) Positive event at end
"I hit five out of six! I've never come anywhere close to hitting the big jackpot before! It was just unbelievable," Cooper exclaimed, still stunned. Berwick bakery worker Tara Cooper stopped off at her usual place for a breakfast coffee and bagel, Brewed Awakening, and decided to pick up a lottery ticket before heading to first shift. "I don't usually play Megabuck$, and don't know why I did today." After work, she checked her numbers online. "I was like, oh my God!"	Berwick bakery worker Tara Cooper stopped off at her usual place for a breakfast coffee and bagel, Brewed Awakening, and decided to pick up a lottery ticket before heading to first shift. **"I hit five out of six! I've never come anywhere close to hitting the big jackpot before. It** was just unbelievable." Cooper exclaimed, still stunned, "I don't usually play Megabuck$, and don't know why I did today." After work, she checked her numbers online. "I was like, oh my God!"	Berwick bakery worker Tara Cooper stopped off at her usual place for a breakfast coffee and bagel, Brewed Awakening, and decided to pick up a lottery ticket before heading to first shift. "I don't usually play Megabuck$, and don't know why I did today." After work, she checked her numbers online. "I was like, oh my God!" Cooper exclaimed, still stunned. "It was just unbelievable. **I've never come anywhere close to hitting the big jackpot before! I hit five out of six!"**

and lead to stronger "lucky" ratings for positively framed scenarios and stronger "unlucky" ratings for negatively framed scenarios.

The third hypothesis we tested regarded the location of the positively or negatively framed information in the long vignettes. In one version of each long vignette, the positively or negatively framed statement was at the start and was followed by background information. In the other two versions, the positively or negatively framed statement was in the middle or at the end of the vignette (see tables 3 and 4). Based on the serial position effect (Rundus 1971), information presented at the beginning (primacy effect) and end (recency effect) of a sequence is remembered better than information presented in the middle of a sequence. Therefore, we expected that the critical positively and negatively framed events would have their greatest impact on perceptions of luckiness when presented at the beginning and end of a long vignette.

A considerable body of research shows that the psychological proximity of facts or information has lasting effects on judgment and choices. One of the best-known examples is that of the availability heuristic, in which information that is especially vivid or memorable outweighs statistical data in decision making (Tversky and Kahneman 1973). People are more afraid of being the victim of a homicide than being the victim of a

TABLE 4. Examples of negatively framed long vignette with critical event moved from beginning to middle to end

Vignette 1 (long version) Negative event at beginning	Vignette 1 (long version) Negative event in middle	Vignette 1 (long version) Negative event at end
"I missed the jackpot by one lousy number! Story of my life. It was just unbelievable," Cooper exclaimed, still stunned. Berwick bakery worker Tara Cooper stopped off at her usual place for a breakfast coffee and bagel, Brewed Awakening, and decided to pick up a lottery ticket before heading to first shift. "I don't usually play Megabuck\$, and don't know why I did today." After work, she checked her numbers online. "I was like, oh my God."	Berwick bakery worker Tara Cooper stopped off at her usual place for a breakfast coffee and bagel, Brewed Awakening, and decided to pick up a lottery ticket before heading to first shift. "I missed the jackpot by one lousy number! Story of my life. It was just unbelievable," Cooper exclaimed, still stunned. "I don't usually play Megabuck\$, and don't know why I did today." After work, she checked her numbers online. "I was like, oh my God."	Berwick bakery worker Tara Cooper stopped off at her usual place for a breakfast coffee and bagel, Brewed Awakening, and decided to pick up a lottery ticket before heading to first shift. "I don't usually play Megabuck\$, and don't know why I did today. After work, she checked her numbers online. "I was like, oh my God." Cooper exclaimed, still stunned, "It was just unbelievable. Story of my life. I missed the jackpot by one lousy number."

suicide, even though (in the United States) the latter is three times as likely. The National Institutes of Health spends more than three times as much to prevent breast cancer in women as it does to prevent prostate cancer in men, even though only 1.3 women die from breast cancer for every man who dies from prostate cancer (National Institutes of Health 2013). The World Health Organization reports that between two hundred and fifty thousand and five hundred thousand people die annually from influenza, and only a few hundred die each year from terrorism (World Health Organization 2003; National Consortium for the Study of Terrorism and Responses to Terrorism [START] 2012). Nevertheless, the fear of terrorism, and the effort to prevent it, far outstrips analogous concern over influenza. The influence of highly visible advocacy groups or frightening and dramatic stories in the media exceeds the power of cold statistics.[1]

Other instances of the serial position effect can lead to counterintuitive and inconsistent preferences, as in the case of the peak-end rule.[2] Robust experimental evidence shows that a person's overall evaluation of a recently ended event (for example, immersing one's hand in cold water,

---

[1] The locus classicus is of course Tversky and Kahneman 1973, but see also the succinct overview of more recent work in Kahneman 2011, esp. chaps. 12 and 13.

[2] A good overview of this research is in Ariely 2008.

a colonoscopy, watching a feel-good television commercial, and receiving gifts) is not determined by the total utility of the experience. Instead, the person's judgment relied on two factors: how good (or bad) the experience was at its peak, and how well (or poorly) the experience ended. One study found that participants preferred sixty seconds of immersion in 14°C ice water followed by thirty seconds of immersion in 15°C ice water to sixty seconds of 14°C ice water alone (Kahneman 2011, chap. 35). Another study found that people retrospectively report lower levels of overall pleasure for a desirable gift if a positive but less desirable gift is added to it, even though the addition of this second gift objectively increases the total worth (Do, Rupert, and Wolford 2008). Nevertheless people will insist that they prefer less pain to more, and more valuable goods to fewer.

Upon testing our hypotheses, we found that the first hypothesis, the existence of a framing effect, was strongly confirmed. Indeed, the framing effect was so powerful that it trumped the length of the vignette; we attained a null result for the second hypothesis. The third hypothesis was interestingly split: where in a vignette the *positive* information was presented did not matter at all in the assessment of its overall luckiness, but the location of *negative* information did matter. In the negative condition, the third hypothesis was confirmed. One possibility is that the lack of a recency effect in the positive vignettes was due to an overall positivity bias among the participants, but we have done no further testing to confirm or disconfirm this additional hypothesis.

2. Method

Participants. 197 students (61 percent female) enrolled in two Introductory Psychology courses were compensated with extra credit to participate in this study, which most of them completed in ten minutes. Most participants were freshmen (76 percent) with a median age of 19 years (age range 18 to 29 years). We chose this sample in order to represent a population of interest naïve to higher-level concepts in Philosophy and Psychology. Based on participants' responses to the sixteen-item Belief in Luck and Luckiness Scale (Thompson and Prendergast 2013), only 9 percent of our sample had "no belief in luck." We hoped to capture laypeople's perceptions of luck. Permission to conduct this research at Bloomsburg University of Pennsylvania was obtained from the local Institutional Review Board. All data were collected anonymously.

Materials. We first created eight short vignettes that could be judged as lucky or unlucky depending on the framing of the scenario (see table 1). A pilot study confirmed our intuition about the perceived luckiness of the vignettes. In all cases, the outcome of the scenario was equivalent. For example, Tara Cooper could **hit five** out of six numbers in a lottery (considered lucky because framed in a positive light) or Tara Cooper could **miss**

one out of six numbers in a lottery (considered unlucky because framed in a negative light). Either way, the actual outcome of the event was the same. Yet we hypothesized that the perceived luckiness of the vignette would change depending upon how the scenario was framed. Those framed in a positive light would be perceived as luckier than those framed in a negative light.

To determine the influence of length of the vignette on perceptions of luck, we created two long versions of each of the eight vignettes, one positively framed and one negatively framed (see table 2). The information included in the positively and the negatively framed vignettes was identical except for the information about the critical event.

To test the primacy-recency effect, we created three versions of each of the positively framed long vignettes (see table 3) and three versions of each of the negatively framed long vignettes (see table 4). Again, all versions of the long vignettes contained the same exact words except for the critical event.

The study was a between-subjects design. Each participant answered only one question about each vignette (eight questions total). However, the vignettes were arranged so that each person experienced each of the eight versions. For example, twenty-five participants completed a survey including the short positive version of Vignette 1, short negative version of Vignette 2, long positive (event at beginning) version of Vignette 3, long positive (event in middle) version of Vignette 4, long positive (event at end) version of Vignette 5, long negative (event at beginning) version of Vignette 6, long negative (event in middle) version of Vignette 7, and long negative (event at end) version of Vignette 8. The next twenty-five participants completed different versions of the eight vignettes, and so on. No participant was presented with both a positive and a negative version of the same vignette, and no participant had both the short and the long version of the same vignette.

Procedure. After providing written informed consent to participate in the study, participants completed three paper-based surveys. The first survey was created by us as described in the Materials section above. Participants were orally instructed to read through each vignette carefully and then make their responses without rereading the vignettes. We hoped to attain participants' first impression of the luckiness of the scenario. Participants indicated their response by circling one of four responses after each vignette: *unlucky, somewhat unlucky, somewhat lucky, lucky.* For example, in Vignette 1, participants were provided with the prompt "Tara Cooper was: *unlucky, somewhat unlucky, somewhat lucky, lucky.* Circle one." In Vignette 2, participants received the prompt "Mark Zabadi was: *unlucky, somewhat unlucky, somewhat lucky, lucky.* Circle one." The instructions in the other vignettes were the same, mutatis mutandis.

After reading and responding to the eight vignettes, participants were instructed to turn over the worksheet to complete the other two surveys.

TABLE 5. Percentages of participants who perceived an event to be "lucky" (includes *lucky* and *somewhat lucky* responses)

Vignette	Short positive	Short negative	Long positive beginning	Long positive middle	Long positive end	Long negative beginning	Long negative middle	Long negative end
1	84%	40%	84%	84%	84%	56%	56%	12%
2	84%	58%	92%	84%	88%	52%	32%	24%
3	68%	4%	54%	48%	80%	12%	8%	8%
4	100%	68%	88%	92%	88%	24%	16%	16%
5	92%	71%	88%	92%	92%	52%	72%	52%
6	72%	16%	96%	92%	80%	38%	9%	16%
7	96%	52%	83%	88%	92%	38%	54%	36%
8	44%	24%	96%	92%	100%	13%	4%	4%

First, participants provided written responses to five demographic questions: age, sex, class year (that is, freshman, sophomore, junior, senior), major in college, and whether English was their first language. After completing the demographic questions, participants completed the sixteen-item Belief in Luck and Luckiness Scale (Thompson and Prendergast 2013). Participants indicated their responses by circling answers, on a scale of 1 (*strongly disagree*) to 5 (*strongly agree*), to questions like "I believe in good and bad luck" and "Belief in luck is completely sensible."

3. Results

Participants' luck ratings (that is, *unlucky, somewhat unlucky, somewhat lucky, lucky*) were tallied individually for each version of each vignette. For ease of interpretation, responses of *unlucky* and *somewhat unlucky* were combined into one "unlucky" category, whereas responses of *lucky* and *somewhat lucky* were combined into one "lucky" category. Percentages of "lucky" responses for all scenarios and vignettes are presented in table 5.

A chi-square test of independence was used to determine if there was a significant influence of framing (that is, positive or negative) on perceptions of luckiness (that is, "unlucky" or "lucky"). The results showed a very strong effect of framing on perceptions of luck, X^2, (2, $n = 1,586$) = 559.6, $p < .001$. When events were framed positively (for example, Tara Cooper **hit five** out of six numbers in the lottery), participants considered the event "lucky" 83 percent of the time. The same events when framed negatively (for example, Tara Cooper **missed five** out of six numbers in the lottery) were considered "lucky" only 29 percent of the time.

Interestingly, when the influence of framing of the event on luck ratings as a function of length of the vignettes was examined, both short and long versions of the positively framed events led to significantly more "lucky" ratings than negatively framed events *(p < .001* in both cases). Negatively framed long vignettes were "unlucky" 71 percent of the time, and negatively framed short vignettes were similarly "unlucky" 72 percent of the time. Positively framed long vignettes were judged "lucky" 86 percent of the time, whereas positively framed short vignettes were judged "lucky" slightly less frequently (78 percent of the time). Overall, the length of the vignette, and thereby the amount of context included in the framing, did not seem to have a significant effect on perceptions of luck. Therefore, framing had significant influences on perceptions of luck regardless of whether the vignette was short or long.

A second chi-square test of independence was used to determine if there was a significant influence of the location of the critical event (that is, placed at the beginning, middle, or end) on perceptions of luckiness (that is, "unlucky" or "lucky") in the long vignettes. The location of the event did not have a significant effect on perceptions of luckiness in the positive long vignettes, X^2 (2, $n = 1,192$)= 1.127, $p = .569$. "Lucky" ratings were fairly consistent whether the events were presented at the beginning (86 percent), in the middle (85 percent), or at the end (88 percent). However, the location of the events did have a significant effect on perceptions of luckiness in the negative long vignettes, X^2 (2, $n = 1,192$)= 10.024, $p = .007$. "Unlucky" ratings seemed to increase as the events were presented closer to the end (beginning = 65 percent, middle = 70 percent, end = 79 percent).

Pritchard and Smith (2004, 24) try to make hay of the skill/external chance division to address earlier work in the psychology of luck. They write, "An outcome that is brought about via an agent's skill is not, we argue, properly understood as a 'lucky' outcome." Similarly, Mauboussin (2012, 24) contrasts luck and skill, and attempts to "place activities properly on the continuum between skill and luck." Since skilled success is typically assumed to be the opposite of luck, we decided to see whether our data supported such a distinction.

Additional chi-square tests of independence were used to determine if there was a significant influence of the type of scenario on perceptions of luckiness. The eight scenarios were first split into categories of skill (Vignettes 2, 4, and 8) and chance (Vignettes 1, 3, 5, 6, and 7). The results showed that framing significantly influenced both types of scenarios: $X^2 skill$(2, $n = 591$) = 208.56, $p < .001$ and $X^2 chance$(2, $n = 995$) = 232.02, $p < .001$. Positively framed scenarios of skill were considered "lucky" 87 percent of the time, and positively framed scenarios of chance were considered "lucky" 83 percent of the time. Negatively framed scenarios of skill were considered "unlucky" 72 percent of the time, and negatively framed scenarios of chance were considered "unlucky" 65 percent of the time. The framing effect swamped any perceptions of difference between skill and chance with respect to luck.

Teigen (2005) claims that luck implies closeness to disaster, and that people judge themselves luckier in cases of a near miss, where they just skirted a terrible outcome, than in cases where they were comfortably separated from trouble. If Teigen is correct, then we should expect that participants would consider near-miss cases to be more a matter of luck than cases where the outcome was half positive and half negative (or the glass is half empty/half full). An example of a near miss is in the Derek Washington vignette, where Derek nearly died in a freak accident with a runaway tractor-trailer tire. An example of a half-empty/half-full scenario is the Mark Zabadi case, where Mark missed half of his basketball free throws but hit half.

To test Teigen's implication, we split the eight scenarios into categories of a near miss (Vignettes 1, 4, 5, and 7) and glass half empty or half full (Vignettes 2, 3, 6, and 8). The results showed that framing significantly influenced both types of scenarios: $X^2 nearmiss\,(2, n = 800) = 171.5, p < .001$ and $X^2 glasshalf(2, n = 786) = 283.4, p < .001$. Positively framed scenarios of near misses were considered "lucky" 89 percent of the time, and positively framed scenarios of glass half empty/half full were considered "lucky" 80 percent of the time. Negatively framed scenarios of near misses were considered "unlucky" only 55 percent of the time, compared to negatively framed scenarios of half empty/half full, which were considered "unlucky" 80 percent of the time. The weaker effect of framing on negatively framed near-miss scenarios is an interesting finding that we will explore in future research. Overall, however, framing had a significant effect on both types of scenarios. Participants' assignments of luck were not notably different in the near-miss cases from what they were in the half-empty/half-full scenarios, contrary to Teigen's prediction.

4. Discussion

Overall, it did not matter whether the salient event in the vignette was ostensibly connected to skill (as in the Mark Zabadi, Vicki Mangano, and José Ramirez examples) or the result of external chance or accident (as in the Tara Cooper, snowstorm, Derek Washington, Michelle Simmons, and James Goldberg examples). Nor did it matter whether the vignette depicted a "near miss" (Tara Cooper, Vicki Mangano, Derek Washington, James Goldberg) or a "the glass is half empty/half full" scenario (Mark Zabadi, snowstorm, Michelle Simmons, José Ramirez). Those factors were not relevant to the luck attributions of the study participants. The statistical determinants of their responses were whether the vignettes were framed negatively or positively and the location of negative information in the vignette.

What is the philosophical import of these results? First, we are not assuming that evidence from empirical psychology immediately refutes or confirms any particular philosophical view. Some overly hasty enthusiasts of experimental philosophy have made this error, and one of us has criticized them in previous work (Hales 2006 and 2012). We do think, however, that the fact that luck attributions are so profoundly subject to cognitive bias—especially framing, but also recency—poses serious challenges to the three prominent theories of luck.

The intuitive responses of naïve study participants rated the same scenarios as lucky or unlucky, depending on whether the information was presented positively or negatively. How can the three major theories of luck, the probability, modal, and control views, explain that fact? One option is to maintain that according to the theories the subjects in each vignette are both objectively lucky and objectively unlucky, and that the study participants correctly perceived this fact. Therefore, the empirical results pose no threat to any theory of luck. A second option is to argue that according to one or more of the theories the subjects in each vignette are either objectively lucky or objectively unlucky (but not both), and that the study participants were simply mistaken when they gave the wrong answer.

Option 1

Consider first the probability theory of luck. According to it, something's luckiness is a function of its importance and probability of occurrence. Recall the Tara Cooper example. Under the probability theory, Tara is lucky to hit five out of six numbers in the Megabuck$ lottery if and only if (1) hitting five numbers mattered to her in a positive way and (2) it was improbable that she would hit five of six numbers. Let us suppose that those conditions were satisfied. Thus Tara *was* lucky to hit five numbers in the lottery. Also according to the probability theory, she was *un*lucky to miss one of six numbers in the Megabuck$ lottery if and only if (1) missing one of the numbers mattered to her in a negative way, and (2) it was improbable that she would hit all six numbers. Assuming those conditions were satisfied, it follows that she was unlucky to miss one of the six numbers.

We get the same result for the modal theory. As Pritchard puts it, "The degree of luck involved varies in line with the modal closeness of the world in which the target event doesn't obtain (but where the initial conditions for that event are kept fixed). We would thus have a *continuum* picture of the luckiness of an event, from very lucky to not (or hardly) lucky at all" (Pritchard 2015). Under the modal theory, a very small change in the world, such as one ball in the Megabuck$ lottery hopper rotating an extra 20 degrees, would have meant that Tara Cooper did not hit five of six numbers in the lottery, and so her hitting those numbers was modally fragile. Thus her success in getting five out of six was lucky. It is also the case that a very small change in the world would have meant that she got all six numbers right in the lottery, and she was unlucky not to find herself in this very close possible world instead. Again, Tara Cooper is both lucky and unlucky for the same thing.

The control theory is no different. The fact that Tara got five of six lottery numbers correct was wholly outside her control. Coupled with the fact that getting those numbers mattered to her, under the control theory she was lucky to get five of six numbers in the lottery. However, it was also not within her control to hit all six numbers, although she would have dearly loved to. Thus the fact that she missed one number was a case of bad luck. While Tara Cooper was lucky to have hit five out of six numbers in the lottery, she was unlucky to have missed one number.

At first, one might think that the fact that all three leading theories of luck give the same result in the Tara Cooper case means that each can nicely explain why participants agreed she was lucky in the **hit five** condition but unlucky in the *missed one* condition. She was both lucky and unlucky, and the study participants correctly recognized this. And of course, similar reasoning applies to the other cases: the town was lucky that half the residents never lost their power in the storm, but unlucky that half did lose their power, and so on.

The problem with the preceding approach is, as Rescher (1995, 212) correctly notes, *being lucky* and *being unlucky* are contrary properties in the same way that *being red all over* and *being blue all over* are contraries, or *skydiving* and *swimming* are contraries. No one can skydive and swim simultaneously, and no one can be both lucky and unlucky for the same thing. One way to see this is to imagine an integer line along which someone's luck might be measured. Such a scale follows directly from Rescher's view, and is consonant with Pritchard's suggestion above that we should adopt a continuum picture of luck.

Unlucky   Lucky

−10 −9 −8 −7 −6 −5 −4 −3 −2 −1 0 1 2 3 4 5 6 7 8 9 10

Someone mildly lucky might have a score of 3 on the scale; someone quite lucky might have a score of 12. The negative integers measure unluckiness. A rating of −4 is fairly unlucky, but −15 is much more unlucky. Zero means that someone is neither lucky nor unlucky; his luck balances out. Sam might be lucky in love but unlucky with investments, which is to say that he scored +5 luck with respect to love and, say, −5 with respect to investments. What is nonsensical is to claim that someone might rate both +5 and −5 on the luck scale with respect to love. One cannot be simultaneously lucky and unlucky with respect to the same thing. Tara Cooper cannot be both positively lucky and negatively unlucky in the lottery case.

The empirical results are that study participants perceived a scenario as lucky when the information was presented in a positive light and unlucky when it was presented in a negative light. At first it seems that all three theories of luck can explain these results in terms of the subjects in the vignettes being both lucky and unlucky. We have just argued, however, that it is not possible for someone to be both lucky and unlucky for the same thing. If the three theories of luck do not fall directly to reductio ad absurdum, at the least they must retool and find some other way of dealing with these results.

A defender of a traditional theory of luck might respond to the preceding criticism as follows. When discussing luck, there is an ambiguity between the lucky/unlucky distinction and the luck/non-luck distinction. While the various accounts of luck are committed to the possibility that the same event can be both good luck and bad luck, that's not a problem for any specific theory of luck at all, since those theories are not an account of good or bad luck, just of whether an event is lucky *simpliciter* (that is, if the subjects agree that the event is good luck/bad luck, then they agree it's lucky, and that's all that's relevant). For our objection to stick we would thus need to disambiguate this distinction along the "P/not-P" lines and then show that the probability, modal, and control accounts are committed to both P and not-P (that is, to the same event being both a case of luck and not a case of luck). Since we have not done so, our criticism goes wide of the mark.[3]

While it is true that there is a valuable distinction between good luck/bad luck and luck/non-luck, that distinction is not germane to our critique. Here's why. Consider an analogy to axiology. There is a difference between the moral/immoral distinction and the moral/non-moral distinction. It is a very plausible assumption that token identical acts cannot be simultaneously moral and immoral; any theory of ethics committed to the same act being both moral and immoral faces a very serious problem.[4] A critic of such a theory need not show that the theory further entails that the very same act is both moral and non-moral. While we have not demonstrated that, say, the modal theory is committed to a person being both lucky and non-lucky for the same thing (which would certainly be bad news for the theory), we *have* shown that it is committed to a person being both lucky and unlucky for the same thing (which is also bad news for the theory).

Another strategy a defender of the traditional theories might pursue is to maintain that the event of **hitting five numbers** is not the same event as *missing one number.* To be sure, such a strategy requires specific commitments in the theory of events. Is Sebastian's stroll at midnight the same event as Sebastian's leisurely stroll at midnight? Is Brutus's stabbing of Caesar the same event as Brutus's killing of Caesar? These matters go back to the well-known Davidson-Kim debate over events.[5] Insisting the positive and negative conditions are really descriptions of separate events in order to preserve the probability, modal, or control theories looks even stranger with some of the other scenarios. For example, the tornado that destroyed half the buildings in town is identical to the tornado that spared half the buildings in town, but that tornado participated in two simultaneous yet

---

[3] Thanks to Duncan Pritchard for this criticism.

[4] As Plato notes at *Euthyphro* 8a–b.

[5] The classic Davidson and Kim pieces are reprinted with commentary in Hales 1999, 319–69. Casati and Varzi 2010 is a good review of recent literature on events.

independent events: destroying half the buildings and sparing half. Still, one might think that the empirical data above simply put a certain amount of metaphysical pressure on any particular theory of luck, not that they are in direct conflict.

Unfortunately, the "different events" response misses its target entirely. As was noted above, the study participants were not asked whether an *event* was lucky/somewhat lucky/somewhat unlucky/unlucky. They were asked whether *a person or an object* was lucky/somewhat lucky/somewhat unlucky/unlucky. Was Tara Cooper, Mark Zabadi, the town, Vicki Mangano, Derek Washington, and so on, lucky or not? Appealing to a certain metaphysics of events will do nothing to accommodate these results. Thus Option 1 is unsuccessful—it cannot be that the vignette subjects were both objectively lucky and objectively unlucky, and to the extent that the theories of luck allow for that possibility, so much the worse for those theories.

Option 2

According to one or more of the theories of luck, the subjects in each vignette are either objectively lucky or objectively unlucky (but not both) and the study participants were simply mistaken when they gave the wrong answer. Given the preceding arguments that every theory of luck seems to give the result that each vignette subject is both lucky and unlucky, we are not clear whether a proponent of, for example, the modal theory would declare that Tara Cooper is objectively lucky or that she is objectively unlucky. Under the present option, she cannot be both. Let's suppose for the sake of argument that she is, in fact, unlucky to have missed one number. In the positive framing case, the study participants are simply misled by the phrasing and wrong to identify her as lucky. (Note that the present option does not address the recency effect at all. No metaphysical account of luck explains why a vignette subject should be judged unluckier as the negative information is moved closer to the end of the vignette.)

The question then becomes one of why we should prefer the *modus ponens* of assuming a theory of luck is right and then dismissing the luck judgments of the study participants over the *modus tollens* of taking those judgments seriously and rejecting any theory of luck that comes into conflict. Philosophical theories of luck aim to solve certain problems in epistemology or ethics or political philosophy, but when they are presented as general metaphysical accounts of what luck is, they must take seriously robust intuitions about luck outside philosophical contexts. The appeal of any theory of luck rests partly upon the degree to which it comports with our pre-analytic intuitions or naïve judgments about luck. To be sure, reflective equilibrium demands that we balance the power and appeal of a theory with data that come into conflict with that theory. Sometimes we do reasonably reject data as outlier, biased, or poorly acquired in favor of preserving an otherwise well-supported theory.

In the present case, though, there is no reason to think that the study participants were laboring under some kind of cognitive illusion, or making a perceptual or inferential error when they judged that Tara Cooper was lucky, but not suffering the same illusions or making the same errors when they judged that she was unlucky. At the very least, a defender of one of the traditional theories of luck needs to provide an error theory that explains not only why naïve participants are wrong in their luck attributions but also why they are systematically, predictably wrong, depending on how the information is presented. We are doubtful that a cogent argument can be made to the conclusion that people make mistakes due to framing and recency effects when the information is negative, but properly grasp the truth when the information is positive. For one thing, sometimes the belief that a subject in a vignette is unlucky will be the correct view, in which case traditional luck theorists will have to hold that judgments that the subject is lucky are due to erroneous bias. The study participants got it right when they thought James Goldberg was lucky, right when they said Michelle Simmons was unlucky, and wrong when they concluded that Derek Washington was lucky? We cannot see how such reasoning might proceed except on a purely ad hoc basis.

5. Conclusion

A reasonable interpretation of our results is that luck is a cognitive illusion and assignments of luck are merely a way to subjectively interpret our experiences; our encounters with the world do not include the detection of a genuine property of luck. Once we see this, it is easy to understand otherwise puzzling claims of luck. Teigen (2005, 129–30) offers this example: "Anat Ben-Tov, survivor of two Tel Aviv bus bomb attacks, expressed this [point] succinctly in an interview, given from her hospital bed: 'I have no luck [i.e. am unlucky] or I have all the luck in the world—I'm not sure which.'" Neither Teigen nor Pritchard (2005, 142), who comments on this case, see its true import. Anat Ben-Tov is unsure how to interpret her involvement with the terrorist bombings without a specific frame to structure that interpretation. As with the vignettes in our study, if her experiences are framed positively as persistent survival in dangerous circumstances, then she has all the luck in the world, but if they are presented negatively as repeated victimization by violent terrorism then she has no luck whatsoever. There is no fact about whether she is objectively lucky or objectively unlucky to be discovered through the application of the probability, modal, or control theory of luck. In our view, philosophers concerned about the role luck plays in philosophical problems are misled about the nature of those problems. An amputee who feels phantom limb pain is misguided about the true source of the pain and requires cognitive therapy to overcome it. While our feelings and intuitions about luck remain even if we recognize them as no more veridical than an itch in a missing leg, we too should see those intuitions as meriting therapy instead of further theoretical analysis.

Acknowledgments

We are grateful to our research assistants José Calvo and Krista Kobbe and also to Mary Katherine Waibel Duncan for helpful comments on a previous version. This work has been funded by Spanish Government (Ministerio de Economia y Competivalid) Research Projects FFI2008-01205, Puntos de Vista. Una Investigación Filosófica, and FFI2011-24549, Puntos de vista y estructuras temporales.

References

Ambegaokar, Vinay. 1996. *Reasoning About Luck: Probability and Its Uses in Physics.* Cambridge: Cambridge University Press.
Ariely, Dan. 2008. *Predictably Irrational.* New York: HarperCollins.
Bewersdorff, Jörg. 2005. *Luck, Logic, and White Lies: The Mathematics of Games.* Wellesley, Mass.: A. K. Peters.
Casati, Roberto, and Achille Varzi. 2010. "Events." In *The Stanford Encyclopedia of Philosophy*, ed. Edward N. Zalta. http://plato.stanford.edu/archives/spr2010/entries/events/.
Do, Amy M., Alexander V. Rupert, and George Wolford. 2008. "Evaluations of Pleasurable Experiences: The Peak-End Rule." *Psychonomic Bulletin and Review* 15, no. 1:96–98.
Greco, John. 2010. *Achieving Knowledge: A Virtue-Theoretic Account of Epistemic Normativity.* Cambridge: Cambridge University Press.
Hales, Steven D. 1999. *Metaphysics: Contemporary Readings.* Belmont, Calif.: Wadsworth.
———. 2006. *Relativism and the Foundations of Philosophy.* Cambridge, Mass.: MIT Press.
———. 2012. "The Faculty of Intuition." *Analytic Philosophy* 53, no. 2:180–207.
Kahneman, Daniel. 2011. *Thinking Fast and Slow.* New York: Farrar, Straus and Giroux.
Kahneman, Daniel, and Amos Tversky. 1984. "Choices, Values, and Frames." *American Psychologist* 39, no. 4:341–50.
Levy, Neil. 2011. *Hard Luck: How Luck Undermines Free Will and Moral Responsibility.* Oxford: Oxford University Press.
Mauboussin, MichaelJ. 2012. *The Success Equation: Untangling Skill and Luck in Business, Sports, and Investing.* Boston: Harvard Business School Press.
Mele, Alfred R. 2006. *Free Will and Luck.* Oxford: Oxford University Press.

National Consortium for the Study of Terrorism and Responses to Terrorism (START). 2012. "Global Terrorism Database." http://www. start.umd.edu/gtd.

National Institutes of Health. 2013. "Estimates of Funding for Various Research, Condition, and Disease Categories." http://report.nih.gov/ categorical_spending.aspx.

Pritchard, Duncan. 2005. *Epistemic Luck*. Oxford: Oxford University Press.

———. 2015. "The Modal Account of Luck." Included in this collection.

Pritchard, Duncan, and Matthew Smith. 2004. "The Psychology and Philosophy of Luck." *New Ideas in Psychology* 22:1–28.

Rescher, Nicholas. 1995. *Luck: The Brilliant Randomness of Everyday Life*. New York: Farrar, Straus and Giroux.

Rundus, Dewey. 1971. "Analysis of Rehearsal Processes in Free Recall." *Journal of Experimental Psychology* 89, no. 1:63–77.

Teigen, Karl Halvor. 2005. "When a Small Difference Makes a Large Difference: Counterfactual Thinking and Luck." In *The Psychology of Counterfactual Thinking*, ed. David R. Mandel, Denis J. Hilton, and Patrizia Catellani, 129–46. London: Routledge.

Thompson, Edmund R., and Gerard P. Prendergast. 2013. "Belief in Luck and Luckiness: Conceptual Clarification and New Measure Validation." *Personality and Individual Differences* 54:501–6.

Tversky, Amos, and Daniel Kahneman. 1973. "Availability: A Heuristic for Judging and Probability." *Cognitive Psychology* 5, no. 2:207–32.

———. 1981. "The Framing of Decisions and the Psychology of Choice." *Science* 211, no. 4481:453–58.

World Health Organization. 2003. "Influenza," fact sheet no. 211. http:// www.who.int/mediacentre/factsheets/2003/fs211/en/.

CHAPTER 4

FRANKFURT IN FAKE BARN COUNTRY

NEIL LEVY

Prior to 1969, the *principle of alternative possibilities* (PAP) was more or less universally accepted. According to PAP, an agent is morally responsible for an action if (inter alia) she could have done otherwise than act as she did. Harry Frankfurt (1969) changed all that when he offered a counterexample to PAP. Since then, philosophers have traded thought experiments and arguments, with many seeking to buttress PAP or to replace it with some successor principle which would capture the highly intuitive thought that responsibility-level freedom requires some kind of two-way power. But very many have conceded the fight; it is now widely held that moral responsibility does not require any such power.

In this essay, I argue that they ought not to have conceded the game. There are good reasons to think that Frankfurt-style cases (FSCs) fail to provide counterexamples to PAP, because there are good reasons to think that the counterfactual interveners who feature in these cases are not appropriately independent of the agent who acts. Their mere presence—despite the fact that, by stipulation, the counterfactual interveners play no causal role at all—affects the normative properties of the agent. This is a conclusion for which I have argued previously (Levy 2008), but many people remain unconvinced (Clarke 2011; Haji and McKenna 2011; Cova 2014; see Levy 2012 and 2014 for replies). Further, the argument I advanced and have defended turns crucially on assessment of agents' responsibility for the consequences of omissions, which makes the argument doubly vulnerable: first, because it may be possible to show that agents are responsible for their volitions, even if they are not responsible for the consequences of their volitions and, second, because omissions cases are notoriously tricky

The Philosophy of Luck, First Edition. Edited by Duncan Pritchard and Lee John Whittington.
Chapters © 2014 Metaphilosophy LLC and John Wiley & Sons Ltd, except for "Luck as Risk and the Lack of Control Account of Luck" © 2015 Metaphilosophy LLC and John Wiley & Sons Ltd. Book compilation © 2015 Metaphilosophy LLC and John Wiley & Sons Ltd.
Published 2015 by John Wiley & Sons Ltd.

and may generate unreliable intuitions. In this essay, I want to take a very different tack.

I develop my argument against FSCs by utilizing resources that come from the debate over epistemic luck. I show that generating the right results in these kinds of cases requires us to concede that (merely) counterfactual interveners affect normative properties. Such interveners—who, once again, never causally interact with the agent—can make the difference between the agent having merely true justified beliefs and having knowledge. I then argue that if counterfactual interveners can make this kind of difference despite their causal inertness, their being disposed to intervene may make a difference to agents' moral responsibility. I conclude that reflection on how counterfactual interveners can make a difference to moral responsibility, apparently by affecting purely modal properties, gives us reason to think that despite the mountains of words spilled on the topic in recent years we still don't properly understand what role alternative possibilities should play in attributions of moral responsibility.

1

In a typical Frankfurt-style case, an agent performs a bad action for which, intuitively, he seems to be morally responsible, despite the fact that he lacks alternative possibilities. In an FSC, the action is one that another agent— canonically Black, a nefarious neuroscientist—wants the first agent (call her Elsie) to perform. Black has implanted a device in Elsie's brain that he will trigger if and only if she seriously considers performing some alternative action; if it is triggered, Elsie's deliberative processes will be bypassed, and she will perform the bad action.[1] As it happens, though, the agent does not seriously consider performing some alternative action, and the device is not triggered. Instead, she performs the action on her own. Because Elsie acts on her own, using her entirely unimpaired capacities for reacting and responding to reasons, she seems to be morally responsible for what she does, but because the device would have caused her to perform the bad action if she had seriously considered an alternative, she lacked alternative possibilities. Hence agents can be morally responsible despite lacking alternative possibilities, and PAP is falsified. At least, that has been the moral that has been widely drawn from these cases.

Now let's transpose the counterfactual intervener, along with his equipment, to fake barn country. In the original fake barn case, an agent—call

---

[1] I borrow from Pereboom (2001) the idea that the neuroscientist's signal for intervention is a thought that is necessary but not sufficient for the agent's deciding in a manner different to the way the neuroscientist wants. If the signal were a sufficient condition for her making the undesired decision, then Elsie might be said to have begun the process of deciding otherwise, and one might worry that she has robust alternatives after all. But if the signal is only a necessary condition, she seems stripped of the ability to decide otherwise, yet still seems to decide on her own, in the ordinary manner.

her Daphne—forms the belief (in the usual manner) that she is standing in front of a barn. Unbeknownst to her, almost all the structures she sees scattered around the countryside are mere facades. By chance, though, Daphne is standing in front of the only genuine barn for miles around. Most epistemologists judge that though Daphne has a true justified belief, she does not know that she is standing in front of a barn. One common way of capturing the thought that her belief does not qualify as knowledge is to say that she is too lucky in getting it right. Were she to have been standing in front of the many barn facades that surround her when she formed her belief it would have been false, and it is merely chance that Daphne happens to be standing in front of the one genuine barn.

The case illustrates the way in which what Duncan Pritchard (2005) calls *veretic epistemic luck* undermines knowledge. A belief is undermined by veretic epistemic luck when it is true in the actual world but false in a wide class of the nearest possible worlds in which the initial conditions are much the same, and the agent forms the belief in the same manner. Daphne could just as easily form her belief *I'm standing in front of a barn* while standing in front of any of the structures she sees around her, almost all of which are mere facades. In most nearby possible worlds, her belief would be false. Her belief is not knowledge, because she is lucky that it is true.

Recall, however, that our counterfactual intervener is standing by. Suppose, now, that Black monitors Daphne as she wanders through fake barn country. He stands ready to use his advanced neuroscientific techniques to cause Daphne to continue to mentally wrestle with the philosophy paper she is working on; he will intervene if she pauses in front of one of the many barn facades and begins to form the belief that she is standing in front of a barn (Daphne is so constituted that it is a necessary, though not sufficient, condition of her forming the belief that she first recalls her early interest in architecture; knowing this, Black monitors her mental activity, ready to trigger thoughts about her paper if she starts to think about architecture while standing in front of a barn facade). Black, who has a phobia with regard to false but justified beliefs due to a traumatic philosophy 101 class, wants Daphne to form the belief that she is in front of a barn only when it is true; hence his readiness to intervene to bring it about that this is the case. As things happen, though, he never needs to deploy his technology: Daphne continues to grapple with her paper until, when she is facing the only genuine barn for many miles around, she recalls her early interest in architecture and goes on to form the belief that she is standing in front of a barn.

Does Daphne *know* that she is standing in front of a barn? Her belief is not subject to veretic epistemic luck. It is not the case that in most nearby possible worlds in which she forms her belief in the same manner as in the actual world, she forms a false belief. So we must conclude either that Daphne knows that she is standing in front of a barn or that Pritchard's principle fails to capture the knowledge-undermining features of all possible fake barn cases.

Which way should we go? Pritchard himself (2012) and Adam Carter (2013) have both recently presented arguments that entail or at any rate might be utilized to show that we ought to deny that Daphne *knows* that she is standing in front of a barn. The arguments raise very different issues; in this section I respond to Carter's, delaying discussion of Pritchard's until the next section.

To construct his case, which he takes to parallel the original fake barn case, Carter adapts a well-known thought experiment from Andy Clark and David Chalmers's (1998) defense of the extended mind hypothesis (Carter's aim is to show that epistemic luck and principles central to extended cognition are incompatible). That thought experiment involved Otto, a sufferer from dementia. Otto's brain-based memory is unreliable; for that reason, he writes in his notebook important facts he knows he may need to recall later (Clark and Chalmers's point, of course, is that Otto's notebook plays the same functional role in his cognitive processes as other agents' biological memories play in theirs, and therefore ought to be recognized as being part of his mind). In Carter's modified version of the Otto case, his notebook has been stolen by a jokester, who changed the times of all his appointments. However, he overlooked one appointment. Otto now consults his notebook regarding that appointment, and forms a true justified belief about its time.

Carter argues that this case is closely analogous to the original fake barn case. The agent in the fake barn case stands by chance in front of the one genuine barn for miles around; the fact that it is by chance that she forms her belief *I'm standing in front of a barn* while in front of the real barn is supposed to prevent her belief from qualifying as knowledge. Similarly, Carter suggests, Otto's belief concerning the time of his appointment is too subject to luck to count as knowledge. Otto's belief is epistemically lucky, Carter claims, because "given how he forms the belief—by consulting a notebook tampered with by a jokester—he could have easily been incorrect" (2013, 4206). How so? As Carter points out, in many nearby worlds the jokester would not have missed the doctor's appointment entry, and therefore Otto would form a false belief in those worlds were he to form his beliefs in the same way as in the actual world.

But notice now that in most nearby possible worlds in which Otto forms his belief in the same manner as in the actual world, he forms a *true* belief. After all, the notebook has the correct time in it. Indeed, it is for this reason that Carter takes his case to be a counterexample to Pritchard's veretic epistemic luck principle: it is false that Otto forms a false belief in a wide class of nearby possible worlds. Though Carter does not say this, it won't help to compare the actual world to a possible world in which the jokester is absent: in such a world, too, Otto consults a notebook with the correct time. So counterfactual error is not nearby, and Pritchard's principle fails to explain why Otto, in the modified case, lacks knowledge. Given the structural similarities between our case involving Daphne and Carter's case involving Otto, if Otto really lacks knowledge it would seem that so does Daphne.

I think that's the wrong result, with regard to both cases. First, modified Otto. This case is not closely analogous to the original fake barn case, I suggest, because Otto's belief is not lucky in the same way in which Daphne's, in the original case, is. In the original fake barn case, Daphne formed her belief by picking out a barn at random; it was therefore mere chance that she picked the only genuine barn. But Otto forms a belief by consulting the same item in his notebook—the one unaltered item—in all (or most) nearby possible worlds. He wants to know the time of his doctor's appointment: he's not simply flipping through the notebook at random. That's not analogous to the fake barn case; at least, it is not analogous to the standard fake barn case. It is more analogous to the following case instead:

Horace forms his belief that he is standing in front of a barn by standing in front of the one genuine barn in fake barn country. Horace selects that particular barn because he is interested in barns with blue doors; all the fake barns have red doors. In almost all nearby possible worlds, Horace would stand in front of the one real barn.

My intuition is that Horace knows that there is a barn in front of him. At very least, we ought to think that his belief is closer to knowledge than is Daphne's in the original fake barn case, since it is the same case with the randomness factor eliminated. Notice, moreover, that this modified case is in one way weaker than the modified Otto case. We are supposing that Otto consults his diary to learn the time of the one unaltered appointment. His belief is counterfactually robust in virtue of his aims as a cognizer (whereas Horace's belief is counterfactually robust in virtue of a quirk of his that has nothing to do with his interest in knowledge). That makes the attribution of knowledge to Otto more certain than the parallel attribution to Horace.

Where has Carter gone wrong? He has failed to locate the luck in the right place, I suggest. Otto is lucky *that this particular diary entry has not been altered.* But given that we hold fixed the diary, with its alterations, and we hold fixed that Otto consults that particular entry, he is not lucky in forming his belief. His luck precedes his formation of the belief. Luck is not transitive: when lucky event e causes event e^*, e^* is not itself lucky in virtue of having been caused by e (I may be lucky to be sitting at my desk today, given that yesterday I had a close shave with a bus, but that doesn't entail that my writing these words is lucky).

So Carter's case is not a counterexample to Pritchard's principle after all, and the way is open for us to treat Daphne, in the presence of the counterfactual intervener, as knowing that she is standing in front of a barn, rather than taking her case to falsify his principle. I think that's the right result (though I shall consider a serious objection shortly). Otto is lucky to know when his doctor's appointment is; he is lucky that the jokester overlooked that entry. But that doesn't entail that his belief is itself lucky. Daphne is not lucky to be standing in front of the one genuine barn in fake barn country when she forms her belief; rather, the counterfactual intervener ensures that she would be there and nowhere else. Unlike the modified Otto case, Daphne's set-up need not involve luck at all: it need not be the case that the counterfactual intervener is present and monitoring her due to luck. We should regard Daphne as akin to Otto in one central way, however. Like Otto, she is not lucky that her belief is true, and like him she should count as knowing what she truly believes.

2

Let's return now to moral responsibility and FSCs. It is highly intuitive that the agents who feature in these cases are morally responsible for what they do. After all, they act on their own, on the basis of their own mental states. The counterfactual intervener does nothing, and how could the mere presence of a counterfactual intervener, who does nothing other than monitor the agent, make a difference to her moral responsibility?

Building on this thought, John Fischer and Mark Ravizza (1998) have argued that in assessing whether agents are morally responsible in FSCs we ought to hold fixed the nonintervention of the counterfactual intervener, asking how the agent would have acted in nearby possible worlds in which (a) she took herself to have a reason to do otherwise and (b) the counterfactual intervener was absent. We should ask about these worlds, rather than worlds in which the counterfactual intervener is present, because these worlds inform us about the agent's capacities: the degree to which she is responsive and reactive to reasons (where an agent is responsive to reasons when she is capable of recognizing them as reasons and reactive to reasons when she would alter her behavior in response to reasons: Fischer and Ravizza have influentially argued that moral responsibility requires a sufficient degree of responsiveness and reactiveness to reasons, including moral reasons). Fischer and Ravizza's test yields the result that agents who feature in FSCs are morally responsible (assuming that they are broadly rational agents in otherwise appropriate circumstances); the counterfactual intervener does not interfere with the properties that underlie moral responsibility.

In brief, Fischer and Ravizza make explicit and give explanatory power to the highly intuitive thought that a merely counterfactual intervener cannot make a difference to whether an agent, who fails even to be aware of the existence of the intervener, is or is not morally responsible. Moral responsibility, it seems, depends on intrinsic properties of agents, not on environmental features that are merely dispositional. But there is a tension between the claim that moral responsibility depends on intrinsic properties of agents alone and our conclusion about the difference that a mere counterfactual intervener can make to whether an agent has knowledge. In the absence of the counterfactual intervener, Daphne's belief *I'm standing in front of a barn* would not qualify as knowledge, but in his presence it does, even though he is as causally inert in the second case as in the first.

To put the point slightly differently, if we hold fixed the nonintervention of the counterfactual intervener in nearby possible worlds in which Daphne forms her belief that she is standing in front of a barn, in order to ask whether her belief qualifies as knowledge, we get the wrong result. If the counterfactual intervener is absent or disabled, then in most nearby possible worlds Daphne forms a false belief. In his absence, her belief, when true, is lucky because it is chancy, and it is chancy because it could so easily have been false.[2] Unless we can identify a principled difference between the domains, such that the epistemic properties of agents can supervene on external features of their environment but their degree of moral responsibility cannot, or we can adduce good grounds for denying that Daphne counts as knowing what she truly believes, the conclusion that merely counterfactual intervention cannot affect moral responsibility seems to be in danger.[3]

Proponents of FSCs will surely urge that we ought to conclude that Daphne does not know that she is standing in front of a barn. They will claim that we ought to go about answering the question whether she counts as knowing what she truly believes in the same way in which, according to Fischer and Ravizza, we should answer the question whether agents in FSCs are morally responsible: by bracketing the counterfactual intervener. That is, we ought to ask whether she would count as knowing what she believes were she to form her belief in the nearest possible world in which Black is absent. Removing him from the scene makes the case identical to the original fake barn case; if this is indeed the right way to go about answering the question, then Daphne does not count as knowing what she believes.

Fischer and Ravizza believe that they are justified in bracketing the counterfactual intervener because it is highly intuitive that agents who feature in FSCs are morally responsible. That the intervener is irrelevant is therefore justified by inference to the best explanation. It is, at very least, less obvious that Daphne does not count as knowing. If this is doubted, note that

---

[2] For chanciness as a necessary, though not sufficient, condition of luck, see Pritchard 2005, Coffman 2007, and Levy 2011. The other conditions that must be satisfied for an event to be lucky, on my view, are absence of control and significance: they are both clearly satisfied in this case.

[3] Actually, there is a third option: accept that Daphne knows in the modified fake barn case, but deny that there is a difference between her epistemic status in this case and the original. Lycan (2006) holds that agents in fake barn cases know what they truly believe, so such a move could be defended. The move would be devastating for my position only if it were accompanied by the denial that there is any difference in the epistemic status of Daphne's belief across cases; the counterfactual intervener need only make *some* difference for my argument to go through.

Daphne's epistemic state *must* be closer to knowledge than the belief of an agent in a standard fake barn case, since the cases are identical except that Daphne is *guaranteed* to have a true belief. Bracketing the counterfactual intervener fails to capture this difference in epistemic status.

That there is such a difference in the status of Daphne's belief, compared to the analogous belief of the agent in the fake barn case, is all I need for my argument.[4] I need only to show that the counterfactual intervener makes a difference, despite his lack of causal impact on Daphne. To that extent, it is not clear that I need to respond to Pritchard's recent supplementation of his account of what it takes for an agent to know something with an ability condition, which Pritchard suggests might entail that Daphne does not count as knowing that she is standing in front of a barn. Even if she does not know, surely the epistemic status of her belief is higher than it would have been in the absence of the counterfactual intervener.[5] Nevertheless, it is worth venturing a few remarks about why I think Pritchard's revised account is compatible with Daphne's having knowledge.

On Pritchard's *anti-luck virtue epistemology,* an agent knows that p if her true belief that p is safe and the product of her cognitive abilities (Pritchard 2012). Daphne's belief is safe: the worry is that she does not count as knowing because her belief is not sufficiently creditable to her cognitive agency. In fact, the case of Daphne in fake barn country seems closely analogous to one of Pritchard's own cases, in which (he suggests) the agent does not count as knowing what he truly believes:

> Temp forms his beliefs about the temperature in the room by consulting a thermometer. His beliefs, so formed, are highly reliable, in that any belief he forms on this basis will always be correct. Moreover, he has no reason for thinking that there is anything amiss with his thermometer. But the thermometer is in fact broken, and is fluctuating randomly within a given range. Unbeknownst to Temp, there is an agent hidden in the room who is in control of the thermostat whose job it is to ensure that every time Temp consults the thermometer the "reading" on the thermometer corresponds to the temperature in the room. (Pritchard 2012, 260).

---

[4] Because I need only to show that the counterfactual intervener makes a difference to the epistemic status of a true belief, my conclusion ought to be acceptable even to those who hold that agents who feature in Gettier cases know what they truly believe. At least this is true if, like Hetherington (surely their most distinguished representative), they accept that there is a difference in the quality of knowledge when it is Gettiered (Hetherington 1999).

[5] Pritchard suggested in correspondence that Daphne's belief might not count as knowledge. If the counterfactual intervener makes a difference to the epistemic status of Daphne's belief—regardless of whether it makes a sufficient difference to make the belief count as knowledge—then that fact has interesting implications for Pritchard's anti-luck virtue epistemology, which identifies an ability condition and an anti-luck condition as necessary and sufficient conditions for knowledge: it would suggest that the satisfaction of each condition independently makes a difference to the epistemic status of a belief, regardless of whether the other is satisfied. Of course, this is a speculation that ought to be tested against a representative range of cases.

Pritchard reports the intuition that Temp doesn't know the temperature, because the correctness of his belief has nothing to do with his abilities. This is a conclusion with which I disagree; I shall return to the point shortly. Right now, however, I want to argue that even if we accept that Temp does not count as knowing what he truly believes, that conclusion does not threaten my claim that Daphne counts as knowing what she truly believes. Indeed, Pritchard himself should accept that the cases are disanalogous. After all, according to him what prevents agents from knowing what they truly believe in fake barn cases is not a failure in their cognitive abilities. On the contrary, for him such cases are counterexamples to traditional virtue epistemology, because in a fake barn case an agent fails to have knowledge *despite* the fact that his "cognitive abilities are putting him in touch with the relevant fact" (Pritchard 2012, 267). It is modal nearness of error, not any problem with cognitive agency, that prevents such agents from having knowledge. It is the nearness of error that the counterfactual intervener eliminates, while (apparently) leaving the agent's abilities untouched. Since the intervener eliminates nearness of error, but it was the nearness of error that would have prevented her from having knowledge, it seems that we ought to conclude that she knows what she truly believes.

What *kind* of difference does the counterfactual intervener make? In earlier work, I argued that one lesson we should draw from FSCs is that agents' abilities may supervene on properties that are partially extrinsic, even on other agents (like the counterfactual intervener).[6] However, the difference between agents who count as knowing what they truly believe and those whose true justified belief fails to count as knowledge need not be a difference in their abilities. The easiest way to see this is to compare the original fake barn case with a case in which Daphne forms her belief in an otherwise identical landscape in which all the barns are genuine. In the second case, but not the first, she counts as knowing, but there is no difference in her abilities across the cases. Rather, there is a difference in her environment that causes her belief to be lucky in the first but not the second. As we have just seen, moreover, the objection that Daphne is like Temp because her cognitive success is not due to her abilities fails; it seems that the counterfactual intervener leaves her abilities unaltered.

Nevertheless, I think the moral of this case and the lesson I drew in earlier work are compatible: counterfactual interveners *can* make a difference to agents' abilities and—given that virtue epistemologists are right in thinking that such abilities play a central role in whether agents count as having knowledge—thereby to their epistemic status. If the counterfactual intervener is a stable enough feature of Daphne's environment, we might want to count the contribution he makes to her epistemic agency as constituting

---

[6] In Levy 2007 and subsequent papers, I talked of agents' *capacities,* rather than *abilities.* I have altered the terminology here, to bring it into line with Pritchard's; so far as I can tell, the difference is merely terminological.

an ability of hers. Consider an agent reliabilist view (Greco 1999). Greco divides those features that underwrite the reliability of a belief-formation processes into those that are part of the agent's "cognitive character" and those that are "strange and fleeting" (1999, 287). However strange they are, counterfactual interveners need not be fleeting; if they play a constant and reliable role, there might be grounds for saying that they help to constitute an agent's cognitive character (think of an external mechanism that intervenes given the appropriate signal). On the other hand, a counterfactual intervener seems to make the difference between Daphne's having knowledge and her having a true justified belief even if his assistance is a one-off event. In that case, we certainly wouldn't want to say he helps constitute part of her character: whether we would nevertheless think he helps constitute some of her abilities depends on how much transitoriness is compatible with the existence of a genuine ability.

Considerations like these help to explain why I don't share Pritchard's view that Temp fails to have knowledge. If the intervention is a stable feature of Temp's environment—which it appears to be, as the case is described—then Temp forms his belief by consulting a thermometer in perfectly good order. Admittedly, the thermometer doesn't work in the way he might imagine it does, since he is ignorant that another agent manipulates it so that its readings match the actual temperature, but ignorance of the workings of the epistemic devices we rely on is ubiquitous. I don't know how my smartphone works, but I seem to know a great many of the facts that I retrieve through it (for that matter, I have only the sketchiest understanding of how my brain works; that doesn't prevent me from gaining knowledge, for instance by retrieving memories). Again, if a property is a stable part of the environment, it might help to constitute an ability (consider a disabled athlete's reliance on her prostheses: there are no grounds to deny that she can run, given that she can outrun most of us).

However we decide on this issue, we should accept that the presence of the counterfactual intervener makes a difference to the normative status of the belief, and—plausibly—of the agent herself (plausibly, since on many views agents are due credit in virtue of bringing it about that they have knowledge; see Pritchard and Turri 2012 for discussion). All by itself that seems to be inconsistent with the conclusion drawn by defenders of FSCs. Bracketing the counterfactual intervener, on the grounds that he is causally inert in the actual sequence, brackets the difference he makes to a normative property: to knowledge. Given that his presence can make a difference to one normative property, there seems every reason to conclude that he is capable of making a difference to another: to the agent's moral responsibility.

Fans of FSCs are wont to ask, rhetorically, how the mere presence of something causally inert could make a difference to an agent's moral responsibility. Concerning one such counterfactual intervener—a device rather than an agent in this case—Fischer asks, "How could the mere existence of such a machine affect the responsibility-grounding relation-ship, given that the machine does not causally interact with the sequence flowing through Jones. ... Indeed, it should be intuitively obvious that the mere existence and operation of the machine ... is *irrelevant* to whatever it is that makes it the case that the responsibility-grounding relationship obtains" (Fischer 2012, 96).[7] At very least, we can see that this rhetorical question loses much of its force: it can no longer be regarded as "obvious" that something that is causally inert is irrelevant to an agent's moral respon-sibility, since we now know that it can affect other normative properties. Nevertheless, Fischer's question remains a good one. How *can* a causally inert agent or machine make a difference to an agent's (or a mental state's) normative properties?

As mentioned, the answer I urged in earlier work is that the counterfac-tual intervener affects the agent's abilities. Very briefly, I want to suggest an alternative now. Let's begin by asking how the counterfactual intervener makes a difference to the epistemic status of Daphne's belief. Let's assume determinism, for ease of exposition, such that it is the case that Daphne was always going to stand in front of the one genuine barn when she formed her belief. The mere fact that she was causally determined to form her belief when it was true does not eliminate luck: luck is compatible with causal determinism (Pritchard 2005; Coffman 2007). Actual access to alternative possibilities is not required for an event to be lucky; the counterfactual pos-sibility of standing in front of one of the many fake barns is sufficient to make a difference to a state's epistemic status. The counterfactual inter-vener eliminates luck by eliminating these counterfactual possibilities. But it is absolutely crucial to defenders of FSCs that the presence of the counter-factual intervener does *not* eliminate alternative possibilities (Fischer 2012, 42). We now have grounds for thinking that causally inert though the inter-vener is, he nevertheless plays precisely this role.

Notice, however, that by eliminating alternative possibilities the coun-terfactual intervener does not appear to bring it about that the agent is not morally responsible for her action. If it is true that counterfactual interven-tion can transform mere true justified belief into knowledge by eliminating luck, then it seems that it ought to be able to *increase* the degree of moral responsibility of agents. Eliminating unactualizable possibilities eliminates luck, and since luck is a factor that reduces moral responsibility, its elim-ination ought to increase agents' degree of responsibility. Transferring the

---

[7] Passages like this seem to entail a commitment to the principle Sartorio calls the "irrel-evance of causally inefficacious factors" (2011, 1072). As she shows, however, that principle is inconsistent with our intuitions concerning switch cases. Sartorio argues that this principle ought to be replaced with a supervenience principle: "There cannot be a difference in respon-sibility for X without a difference in the actual causal sequence issuing in X" (2011, 1073). But if the argument put forward in this essay succeeds, the supervenience principle falls alongside the principle regarding causally inefficacious factors: the counterfactual intervenes in my cases do not affect what is the cause of what, yet they do make a normative difference.

moral of Daphne's case to the free will debate therefore seems to show that defenders of FSCs were right all along in thinking that the agents who feature in these cases are morally responsible, but wrong in thinking that this was because alternative possibilities are irrelevant to moral responsibility. Rather, the agents are *more* morally responsible *because* alternative possibilities are relevant to moral responsibility.[8]

In the modified fake barn case, Black raises the epistemic status of Daphne's belief by eliminating the effects of luck. He can do so only because he eliminates counterfactual error. In the standard FSC, he can be said to have a parallel influence: eliminating counterfactual possibilities. But if that's the case, then we are not entitled to say that his presence is, as defenders of FSCs claim, *irrelevant* to moral responsibility. And the entire verdict, that FSCs show that we can be morally responsible despite lacking alternative possibilities, is cast into doubt. What verdict we ought to draw from these cases concerning the precise role of alternative possibilities is unclear. Are the alternative possibilities eliminated by counterfactual interveners of a different kind from those attention to which motivated the principle of alternative possibilities in the first place, making a different kind of difference to agents' moral responsibility? This is a question I do not think we can answer at present.

I have not aimed to settle what kind of difference (if any) alternative possibilities make to agents' moral responsibility. My aim, instead, has been negative: to show that we are not entitled to the intuitions that FSCs provoke in us. When we consider the roles that counterfactual interveners play in other scenarios, our belief that they make no difference, because they do not causally interact with the agents who feature in them, ought to be shaken. Counterfactual interveners, it turns out, may remain counterfactual while nevertheless affecting normative properties. Defenders of FSCs therefore must offer us some rationale for thinking that they do not play such a role in their favorite cases.

Acknowledgments

Work leading to the completion of this essay was funded by the Australian Research Council. I am grateful to Duncan Pritchard and Lee Whittington for helpful comments on an earlier draft.

---

[8] If counterfactual interveners can increase the degree to which some normative property applies to a case or an agent, can they also reduce it? It seems so. In *rare fake barn country,* almost every barn is genuine. Gloria forms her belief *That's a barn* while standing in front of one of the genuine barns. Unbeknownst to her, however, a counterfactual intervener was standing by. He rolled a dice when he detected that Gloria had satisfied a necessary but not sufficient condition for forming the belief; if he had rolled an even number, he would have intervened to cause Gloria to move on to the fake barn. Gloria would have had knowledge in his absence, but his presence ensures that her belief was too lucky to count as knowledge. If counterfactual intervention can downgrade knowledge to justified true belief, perhaps it can also sufficiently decrease an agent's degree of moral responsibility such that she is not blameworthy for an action.

References

Carter, J. A. 2013. "Extended Cognition and Epistemic Luck." *Synthese* 190:4201–14.

Clark, A., and D. Chalmers. 1998. "The Extended Mind." *Analysis* 58:7–19.

Clarke, R. 2011. "Responsibility, Mechanisms, and Capacities." *Modern Schoolman* 88:161–69.

Coffman, E. J. 2007. "Thinking About Luck." *Synthese* 158:385–98.

Cova, F. 2014. "Frankfurt-Style Cases User Manual: Why Frankfurt-Style Enabling Cases Do Not Necessitate Tech Support." *Ethical Theory and Moral Practice* 17:505–21.

Fischer, J. M. 2012. *Deep Control: Essays on Free Will and Value.* Oxford: Oxford University Press.

Fischer, J. M., and M. Ravizza. 1998. *Responsibility and Control: An Essay on Moral Responsibility.* Cambridge: Cambridge University Press.

Frankfurt, H. 1969. "Alternate Possibilities and Moral Responsibility." *Journal of Philosophy* 66:829–39.

Greco, J. 1999. "Agent Reliabilism." *Philosophical Perspectives* 13:273–96.

Haji, I., and M. McKenna. 2011. "Disenabling Levy's Frankfurt-Style Enabling Cases." *Pacific Philosophical Quarterly* 92:400–14.

Hetherington, S. 1999. "Knowing Failably." *Journal of Philosophy* 96:565–87.

Levy, N. 2008. "Counterfactual Intervention and Agents' Capacities." *Journal of Philosophy* 105:223–39.

———. 2011. *Hard Luck: How Luck Undermines Free Will and Moral Responsibility.* Oxford: Oxford University Press.

———. 2012. "Capacities and Counterfactuals: A Reply to Haji and McKenna." *Dialectica* 6:607–20.

———. 2014. "Countering Cova: Frankfurt-Style Cases Are Still Broken." *Ethical Theory and Moral Practice* 17:523–27.

Lycan, W. G. 2006. "On the Gettier Problem." In *Epistemology Futures*, ed. S. Hetherington, 148–68. Oxford: Oxford University Press.

Pereboom, D. 2001. *Living Without Free Will.* Cambridge: Cambridge University Press.

Pritchard, D. 2005. *Epistemic Luck.* Oxford: Oxford University Press.

———. 2012. "Anti-luck Virtue Epistemology." *Journal of Philosophy* 109:247–79.

Pritchard, D., and J. Turri. 2012. "The Value of Knowledge." *Stanford Encyclopedia of Philosophy.* http://plato.stanford.edu/entries/knowledge-value.

Sartorio, C. 2011. "Actuality and Responsibility." *Mind* 120:1071–97.

CHAPTER 5

LUCK AND FREE WILL

ALFRED R. MELE

In *Free Will and Luck* (Mele 2006), I posed a problem about luck for libertarians (theorists who hold that free will is incompatible with determinism and that there are human beings with free will) and I offered a solution. In this essay, I sketch the problem and examine some recent reactions to it.

1. Some Background

Consider the following case (from Mele 2006, 25). Intending to vote for Gore, Al pulls the Gore lever in a Florida voting booth. Unbeknownst to Al, that lever is attached to an indeterministic randomizing device: pulling it gives him only a 0.001 chance of actually voting for Gore—that is, of actually registering a Gore vote. Luckily, he succeeds in voting for Gore. Beyond the rigging of the voting booths at Al's voting establishment, there is no monkey business in Al's story. He is not brainwashed, for example. And Al is a sane, rational adult whose intention is backed by reasons he had for voting for Gore.

Did Al freely vote for Gore? Was his voting for Gore a free action? Did he vote for Gore of his own free will? These are three ways of asking the same question (at least as I use the relevant terms). If it is assumed that free actions are common in Al's world, then, I submit, the intuitive answer is yes.

What may be said in support of the yes answer, given the assumption I mentioned? If we understand trying to *A* in an unexacting way that is popular in the philosophy of action literature, we should say that Al tried to vote for Gore. Trying to *A*, on the conception of it at issue, requires no *special* effort. For example, when I turned my computer on this morning, I

The Philosophy of Luck, First Edition. Edited by Duncan Pritchard and Lee John Whittington.
Chapters © 2014 Metaphilosophy LLC and John Wiley & Sons Ltd, except for "Luck as Risk and the Lack of Control Account of Luck" © 2015 Metaphilosophy LLC and John Wiley & Sons Ltd. Book compilation © 2015 Metaphilosophy LLC and John Wiley & Sons Ltd. Published 2015 by John Wiley & Sons Ltd.

tried to turn it on, even though I turned it on simply by pressing a button. I expended very little energy and very little effort, but trying to turn on my computer does not require much of either. Now, if free actions are common in Al's world, then it is plausible that he freely *tried* to vote for Gore, given the details of the case. And Al's voting for Gore may count as a free action in virtue of its relationship to his free, successful attempt to vote for Gore. If his voting for Gore is properly counted as a free action on these grounds, then his voting for Gore may be said to derive its status as a free action at least partly from a free action in which his trying to vote for Gore partly consists—that is, from his pulling the "Gore" lever. We can say that his pulling the Gore lever is a directly free action and that his voting for Gore is indirectly or derivatively free.

A note of caution about action individuation is in order. I take no stand on competing theories of the matter. I mention just two such theories here, but I hope that what I say about them will forestall confusions that result from mixing modes of action individuation in one's thinking about cases. On a fine-grained view, Al's pulling the Gore lever is a different action from his voting for Gore. On a coarse-grained view, they are the same action under different descriptions. Just as, on the latter view, the same action can be done intentionally (and for reason R) under one description and unintentionally (and not for reason R) under another, this view leaves it open that the same action may be directly free under one description and indirectly free under another. Readers should understand my action variable "A" as a variable for actions themselves or actions under descriptions, depending on their preferred theory of action individuation. The same goes for the expressions that take the place of "A" in concrete examples. For example, fans of the coarse-grained theory should read "Al pulls the Gore lever" as "something Al does under the description 'pulls the Gore lever,' " and fans of other theories should make no adjustments. Readers who opt for the coarse-grained theory should also read "Al's voting for Gore was indirectly free" as "under the description 'voting for Gore' what Al did was indirectly free," and, again, other readers should make no adjustments. (For more on action individuation in a related context, see Mele 2010, 102–3.)

Deciding to vote for Gore is no part of my story about Al. Decisions to do things, as I understand them, are responses to uncertainty about what to do (Mele 2003, chap. 9). If Al was at no point uncertain about whether to vote for Gore (if voting for Gore was, as we say, a no-brainer for him all along), then there is no place in Al's story for his making a decision to vote for Gore. But imagine another voter, Ann, who is uncertain about whom to vote for (and even about whether to vote at all) and eventually decides to vote for Gore. Ann votes at the same place as Al, has the same chance of actually producing a Gore vote by pulling the Gore lever, and, like Al, luckily succeeds in voting for Gore. Also, like Al, she is rational, unmanipulated, and so on. If Ann freely votes for Gore, her voting for Gore may be said to derive its status as a free action at least partly from her freely *deciding* to vote for Gore. Her trying to vote for Gore may also be said to derive its status as free from her freely deciding to vote for him.

Some philosophers take moral responsibility and the freedom most closely associated with it "to apply primarily to decisions" (Pereboom 2001, xxi). Many philosophers have asserted or argued that to decide to A is to perform a mental action of a certain kind—an action of forming an intention to A (see Frankfurt 1988, 174–76; Kane 1996, 24; Kaufman 1966, 34; McCann 1986, 254–55; Mele 1992, 156, and 2003, chap. 9; Pink 1996, 3; and Searle 2001, 94). I have defended the view that deciding is a momentary mental action of intention formation that resolves uncertainty about what to do (Mele 2003, chap. 9). In saying that deciding is momentary, I mean to distinguish it from, for example, a combination of deliberating and deciding. A student who is speaking loosely may say, "I spent hours last night deciding to major in Philosophy" when what he means is that he spent hours deliberating or fretting about what major to declare and eventually decided to major in Philosophy. Deciding to A, as I conceive of it, is not an extended process but a momentary mental action of forming an intention to A, "form" being understood as an action verb.

In Mele 2005 and 2006, I posed a problem about deciding for libertarianism that is framed partly in terms of luck.[1] I have seen assertions that I offered an argument against libertarianism, but those assertions are false. In fact, I developed a solution of my own to the problem (2006, chap. 5), and the solution was not a rebuttal of an argument. Regarding what I called "the problem of present luck" (2006, 66), I wrote: "My aim in developing this chapter's central problem for agent causationists and other conventional libertarians is to present it sufficiently forcefully to motivate them to work out solutions to it—proposed solutions that I and others can then assess" (2006, 70; see Mele 2005, 414).

So what is the problem? I have devoted a lot of ink to making the problem of present luck salient (2005 and 2006, 5–9, chap. 3, chap. 5)—way too much to permit a thorough recap. Here I briefly sketch the problem.

Consider the following story (from Mele 2006, 73–74). Bob lives in a town in which people make many strange bets, including bets on whether the opening coin toss for football games will occur on time. After Bob agreed to toss a coin at noon to start a high school football game, Carl, a notorious gambler, offered him $50 to wait until 12:02 to toss it. Bob was uncertain about what to do, and he was still struggling with his dilemma as noon approached. Although he was tempted by the $50, he also had moral qualms about helping Carl cheat people out of their money. He judged it best on the whole to do what he had agreed to do. Even so, at noon, he

---

[1] In Mele 2005 (p. 412) and 2006 (p. 70), I say that if "luck" is not the best short label for what I have in mind, I am open to correction. The problem exists no matter what we call it. So does an acceptable solution, I hope.

decided to toss the coin at 12:02 and to pretend to be searching for it in his pockets in the meantime.

According to typical libertarian views (setting aside derivatively or indirectly free actions and actions for which an agent is derivatively or indirectly morally responsible), Bob freely makes his decision and is morally responsible for making it only if there is another possible world with the same past up to noon and the same laws of nature in which, at noon, Bob does not decide to toss the coin at 12:02 and does something else instead. In some such worlds, Bob decides at noon to toss the coin straightaway. In others, he is still thinking at noon about what to do. There are lots of other candidates for apparent possibilities: at noon, Bob decides to hold on to the coin and to begin singing "Purple Haze" straightaway; at noon, Bob decides to start barking straightaway while holding on to the coin; and so on. The "candidates for apparent possibilities" are genuine possibilities if Bob's doing these things at noon is compatible with the actual world's past up to t and its laws of nature. The genuine possibilities are, as I put it in a recent paper (where I avoid putting things in terms of luck), different possible *continuations* of a (normally very long) world segment (Mele 2013).

In the same paper (Mele 2013), I invite my readers to imagine a genuinely indeterministic number generator. At five-minute intervals, consistently with the past up to the pertinent time and the laws of nature, it can generate any one of many numbers or no number at all. Its generating the number 17 at t is one possible continuation of things, and the same is true of many other numbers. At noon today, the machine generated the number 31. After you verify that, you might find yourself with the following belief: the machine's generating the number 31 was a possible continuation of the past up to noon, and that continuation actually happened at noon.

If you were somehow to verify that, at noon, Bob decided to toss the coin at 12:02 and to pretend to be searching for it in his pockets in the meantime (decided to C, for short), you might find yourself with a parallel belief: Bob's deciding to C was a possible continuation of the past up to noon, and that continuation actually happened at noon. As I mentioned, typical libertarians contend that Bob's being directly morally responsible for deciding to C and his directly freely deciding to C require that at least one other continuation was possible at noon, a continuation in which Bob does something else at noon.[2] Suppose that another possible continuation was Bob's deciding at noon to toss the coin straightaway; in another possible world with the same past as the actual world up to t and the same laws of nature, that is what happens.

This supposition may be seen by some as a double-edged sword. A philosopher may believe that having *control over whether* one A-s or does

---

[2] Bob's story is not a Frankfurt-style story. On a version of the continuation problem for fans of Frankfurt-style stories, see Mele 2013, 247–48.

something else instead is required for directly freely A-ing and for being directly morally responsible for A-ing and believe that having such control requires that A-ing at t and doing something else instead at t are possible continuations of the past up to t for the agent. And the very same philosopher may worry that these possible continuations are similar enough to possible continuations for the indeterministic number generator that whatever control the agent may have over whether he A-s or does something else instead falls short of what is required for directly free A-ing and for direct moral responsibility for A-ing.

Consider a fuller version of Bob's story in which although, in world $W1$, Bob does his very best to talk himself into doing the right thing and to bring it about that he does not succumb to temptation, he decides at noon to C. Imagine as well that in another possible world, $W2$, with the same past up to noon and the same laws of nature, Bob's best was good enough: he decides at noon to toss the coin straightaway. That things can turn out so differently at t (morally or evaluatively speaking) despite the fact that the worlds share the same past up to t and the same laws of nature will suggest to some readers that Bob lacks sufficient control over whether he makes the bad decision or does something else instead to make that decision freely and to be morally responsible for the decision he actually makes (again, it is the direct versions of free action and moral responsibility that are at issue). After all, in doing his best, Bob did the best he could do to maximize the probability (before t) that he would decide to do the right thing, and, even so, he decided to cheat. One may worry that what Bob decides is not sufficiently "up to him" for Bob to be directly morally responsible for making the decision he makes and for it to be a directly free decision.

Given the details of Bob's story, how can Bob have enough control over whether he decides to C or does something else instead at noon for his decision to be directly free and for him to be directly morally responsible for deciding to C? This is an instance of the central question posed by what I called "the problem of present luck" (2005, 411, and 2006, 66).

The difference at noon between worlds $W1$ and $W2$ seems to be just a matter of a difference in luck—in which case, each decision seems to be partly a matter of luck. I do not claim that the luck involved is incompatible with Bob's deciding freely. Instead, I ask for an explanation of their compatibility.

I have heard it said that what I am presenting as a problem for typical libertarians cannot possibly be a problem for them because their view entails that cross-world differences of the sort at issue are required for directly free action. But, of course, sometimes a philosopher's view entails something impossible. The question how or why directly free action is possible in a story like Bob's is a fair question. And the answer that it has to be possible because its possibility is required by typical libertarian views is a remarkably poor answer.

It is time to link Bob to the Florida voters. How, someone may ask, can I happily say that Al and Ann freely voted for Gore despite the great luck involved in their succeeding in voting for him and yet worry about whether Bob freely decided to cheat? Two points should be made in response. First, one must bear in mind an explicit assumption in my discussion of the voters. For example, I wrote: "If it is assumed that free actions are common in Al's world, then, I submit, the intuitive answer [to the question whether he freely voted for Gore] is yes." If typical libertarians are right about what deciding freely requires, the problem of present luck apparently threatens that assumption. Second, whereas if Al and Ann freely voted for Gore, the status of those actions as free derives from other free actions, Bob's decision to cheat is supposed to be directly free. Until we understand how directly free actions (including decisions) are possible, we will not be in a position to see how indirectly free actions are possible.

2. Luck and Agent Causation

Some philosophers view the problem of present luck (or something very similar) as a problem for event-causal libertarians but not for agent causationists. Derk Pereboom argues that on event-causal libertarian views, alleged free choices are "partially random" events (2001, 54) in the sense that "factors beyond the agent's control [nondeterministically] contribute to their production ... [and] there is nothing that supplements the contribution of these factors to produce the events" (48). Similarly, Timothy O'Connor refers to "a chancy element to choice that cannot be attributed to the person" in a representative event-causal libertarian view, and he deems "the kind of control that is exercised . . . too weak to ground [the agent's] responsibility for which of the causal possibilities is realized" (2000, 40). O'Connor contends that an upshot of typical event-causal libertarian views is that agents do not directly control what they choose: "There are objective probabilities corresponding to each of the [possible choices], but within those fixed parameters, which choice occurs on a given occasion seems, as far as the agent's direct control goes, a matter of chance" (2000, xiii; see 29). Both Pereboom and O'Connor look to agent causation for a solution to the problem they have in mind, a matter that I have discussed elsewhere (Mele 2006, chap. 3).

In an article I have not discussed elsewhere (Griffith 2010), Meghan Griffith follows suit. Central to her reply to the problem of present luck is the following claim, which she dubs RUL: "In the free will context, something is a matter of luck for A if and only if it happens to A" (2010, 46). Griffith writes: "RUL cashes in on the distinction between doing and happening. What happens to someone is not something she does. In the free will context, that something happens to A is sufficient for its being a matter of luck, since it rules out her having done it. Since this is ruled out, her responsibility for it is undermined" (2010, 46). In Griffith's view, decisions involving exercises of agent-causal power do not happen to the agents and therefore are not even partly a matter of luck.

As the reader will have surmised, I believe that RUL is false. Distinguish trying to vote for Gore from voting for Gore. Al does both in my story. His doing the latter was partly a matter of luck. But it was not something that happened to him. Al's voting for Gore was an action—and a free action, I suggested, if free actions are common in Al's world.

Here is another example. Ann is a hockey player. She takes a shot at the goal. The puck is veering a bit wide when it ricochets off a defender and bounces into the goal. Ann scored a goal. That's something she did, not merely something that happened to her. And her scoring the goal involved some luck; it was partly a matter of luck. Furthermore, if Ann is a normal, rational, unmanipulated agent in normal circumstances and free actions are common in her world, it is plausible that her scoring the goal was a free action.

If Al's voting for Gore and Ann's scoring the goal are free actions, they would seem to be indirectly free. Griffith may seek to avoid the problem I identified for RUL by modifying it as follows: (RUL*) In the context of directly free actions, something is a matter of luck for A if and only if it happens to A. Now, I claimed that Bob's deciding to cheat is partly a matter of luck. In my view, again, to decide to do something is to perform an action of a certain kind. And, on Griffith's view, the very fact that Bob performed the action of deciding to cheat is incompatible with that action's being partly a matter of luck. In her view, apparently, x is not even partly a matter of luck for a person unless x happens to that person; and, as I have mentioned, she asserts that the fact that x happens to a person "rules out her having done it" (2010, 46).

Griffith's view does not fare well when it comes to such actions as Al's voting for Gore and Ann's scoring the goal. They are luck-involving actions—actions that are partly a matter of luck. Does it fare any better in the case of Bob's deciding to cheat?

Griffith says that when we take an event-causal libertarian perspective on an agent faced with a pair of options, A and B, we see that "it just happens to her that the decision is to A rather than to B because she is missing the power to determine the decision" (2010, 51). When Bob is viewed from this perspective, Griffith presumably would say that it just happens to him that his decision is to cheat rather than to flip the coin. And, guided by RUL (or RUL*), she would deduce that the fact that his decision is to cheat rather than to flip the coin is a matter of luck. But if that fact is a matter of luck, what recommends the view that his deciding to cheat is not even partly a matter of luck? Griffith can claim that it cannot be even partly a matter of luck because it is an action and therefore is not something that happens to Bob. But this line of defense is unsatisfactory, as I have explained. Ann's scoring the goal is an action *and* partly a matter of luck, and the same is true of Al's voting for Gore.

In Griffith's view, the event-causal libertarian perspective leaves out something that is required for free will—namely, agent causation. Agents, as characterized by event-causal libertarians, are said to be "missing the power to determine the decision" (Griffith 2010, 51). What power is that? What does it amount to? What is it for an agent to *determine* a decision? Griffith asserts that an agent who lacks the agent-causal power "seems not to have control over the crucial element for which she is responsible: that she has decided to A rather than to B" (2010, 50). She compares this agent to a man in a story by Robert Kane who tries to smash a glass table by hitting it with his arm (Kane 1999, 227). The table broke, but it was undetermined whether the man's striking it would break it. Griffith writes: "Although [he] causes the table's breaking, he does not *completely control* whether the table breaks. In this sense, its breaking happens to him" (2010, 50, emphasis altered). Taking our lead from this, we have an answer to my questions about the alleged power to "determine the decision." It is the power to *completely control* which decision one makes.

What is it about agent causation that underwrites the claim that in scenarios like Bob's (some) agent-causes have the power to "completely control" which decision they make? A good answer to that question might enable me to see that and why the problem of present luck is an illusion—or a problem only for libertarian views that make no use of agent causation. I have not yet seen such an answer.

In my view, even if the difference between what an agent does at t in one world and what he does at t in another world with the same past up to t and the same laws of nature is just a matter of luck, the agent may act freely at t in both worlds (Mele 2006, chap. 5). I dub this thesis *LUCK*. Partly because I accept *LUCK*, I do not search for a notion of control that allows for "complete control" over what one decides to be exercised at t in one or both worlds.

As I see it, if the pertinent difference at t between a world in which an agent decides at t to A and a world with the same past up to t and the same laws of nature in which he decides at t to B is just a matter of luck, then he does not exercise complete control over whether he decides at t to A or instead decides at t to B. Even so, if *LUCK* is true, he may make these decisions freely. By "complete control" I do not mean "as much control as metaphysically possible." I leave it open that the following conjunction is true: exercising the power of agent causation is required for exercising complete control over whether one decides at t to A or instead decides at t to B in scenarios of the sort at issue, and agent causation—conceived of in such a way as to allow for an agent's having complete control over what he decides in scenarios of the sort at issue—is metaphysically impossible.

O'Connor (2011, 324–25) quotes the following from Mele 2006, 70: "If the question why an agent exercised his agent-causal power at t in deciding to *A* rather than exercising it at *t* in any of the alternative ways he does in other possible worlds with the same past and laws of nature is, in principle, unanswerable ... because there is no fact or truth to be reported in a correct answer ... and his exercising it at *t* in so deciding has an effect on how his life goes, I count that as luck for the agent." O'Connor then writes: "Suppose we take this as a stipulative account (or sufficient condition) on luck 'as Mele understands the notion.' If so, it is open to the agent causationist to deny that luck in this stipulated sense is of any significance whatsoever— not, for example, being relevant to freedom and moral responsibility. Mele in fact agrees! He does not press his luck objection as a deep skeptical worry about indeterministic freedom. ... Instead, he wields it to neutralize the agent causationist's objection to causal indeterminism" (2011, 325).

My response is yes and no. Yes, I have never appealed to luck in an argument for the falsity of libertarianism. In fact, I have never argued for the falsity of libertarianism. But no, I do not agree that luck in my sense is of no significance in the spheres of free will and moral responsibility, and no, my aim in articulating my worry about present luck was not to neutralize "the agent causationist's objection" to event-causal libertarianism. My aim was to motivate libertarians of all kinds to produce solutions to the problem of present luck. A welcome solution would explain why the presence of present luck in a case like Bob's is compatible with Bob's directly freely deciding to cheat and with his being directly morally responsible for deciding to cheat. As I observed elsewhere, I regard my central question as an analogue of a request for a theodicy in response to the problem of evil—an explanation of why a perfect God would allow all the pain and suffering that exists and, of course, has ever existed and will ever exist (2013, 241–42). A philosopher who offers an argument from evil for the nonexistence of God should expect rebuttals of the argument. A philosopher who presents the problem of evil as vividly as she can and then asks for an explanation of why a perfect God would allow all the pain and suffering at issue may hope to receive an attractive explanation. For my own analogue of a theodicy in the spheres of moral responsibility and free will, see Mele 2006, chap. 5. (It makes no appeal to agent causation.)

O'Connor reports that in reply to my worry about present luck, "[Randolph] Clarke (2005) argues that an agent causal capacity would provide a stronger variety of control than is available on causal indeterminism" (2011, 325). Be this as it may, Clarke reports that, in his judgment, relevant arguments collectively "incline the balance against the possibility of substance causation in general and agent causation in particular" (2003, 209). Clarke argues (2003) that agent-causal powers are required for free will (at least, if incompatibilism is true). If agent causation is required for free will and impossible, free will is impossible. Recall the reference to a double-edged sword in section 1.

As I have observed, some agent causationists regard the problem of present luck or something very similar as a decisive problem for

event-causal libertarianism. And yet, some of the same agent causationists seem to regard the problem as no threat at all to agent-causal libertarianism. Why might that be? Consider the following from O'Connor: "The
agent causationist takes it to be a virtue of her theory that it enables her
to avoid a 'problem of luck' facing other indeterministic accounts. Agent
causation is precisely the power to directly determine which of several possibilities is realized on a given occasion" (2011, 325). This claim may be
combined with the idea, mentioned earlier, that this determining power
is the power to completely control which decision one makes to yield the
following assertion: Agent causation is precisely the power to completely
control which decision one makes.

Now, anyone can *say* that something or other is the power to completely control which decision one makes. An event-causal libertarian can
say this about some non-agent-causal decision-making power, and so can
a noncausalist libertarian. One thing I would like to know is how replacing
event-caused decisions or uncaused decisions with agent-caused decisions
(or intentions) is supposed to make true something that would otherwise
supposedly be false—namely, that the decisions are made by someone who
exercised the power to completely control which decision he made (or which
intention he came to have). (On this sort of thing, see Mele 2006, chap. 3.) A
plausible answer would be a plausible reply to the problem of present luck.
Until I see such an answer, I will take comfort in the thesis I called *LUCK*.

O'Connor notes that Richard Taylor (1966) "propounded agent causation as a feature of all intentional action" (2011, 311). It is implausible (to
put it mildly) that in the case of every intentional action, agents have and
exercise complete control over whether they perform that action. When
a professional basketball player with a 90-percent success rate at sinking
free throws sinks a free throw in the normal (for him) way, he intentionally
sinks it. But he would seem not to have complete control over whether
he sinks it. A free throw shooter who has and always exercises complete
control over whether he sinks his free throws sinks every free throw he
tries to sink. So, as at least some agent causationists think of agent causation, the agent-causal power to A seemingly does not include complete
control over whether one A-s as an essential feature (for all A-s). And if
having "the power to directly determine which of several possibilities is
realized on a given occasion" (O'Connor 2011, 325) necessarily includes
having complete control over which of the possibilities is realized on that
occasion, then, as some agent causationists conceive of agent causation,
the agent-causal power to A seemingly does not always include the power
to directly determine that one A-s.

A few paragraphs ago, I asked what it is about agent causation that
underwrites the claim that in scenarios like Bob's (some) agent-causes have
the power to "completely control" which decision they make. One option
for someone who conceives of agent causation differently from the way
Taylor does is simply to stipulate that agent causation is the power to

completely control which decision one makes or which intention one comes to have in situations of the sort at issue. But the stipulation is unilluminating.

What is complete control? Considering this question in the context of a simple real-world experiment will prove useful. Only part of the experiment is relevant for my purposes. That is the part I will describe.

Subjects are instructed to press either the Q key on a computer keyboard with their left index finger or the P key with their right index finger (and never to press both at the same time). They are told that which key they press is up to them. They will make more than forty key presses—either Q or P, sometimes one and sometimes the other—in the course of an hour, and they are asked to refrain from planning in advance which key to press. Before they press, they will hear a tone that they are instructed to treat as a "decide" signal. When they hear the tone, they are to decide right then which key to press and then press it straightaway. Pressing a key, as defined for the purposes of this experiment, requires that the key move all the way down and make contact with the switch under it.

What would it be for Sam to have complete control over whether he presses Q or P when he hears the next "decide" signal? Consider the following suggestion. For Sam to have complete control over this is for the following things to be true: at the relevant time Sam is able to try to press Q and able to try to press P, and, regarding each key, if he tries to press it there is no chance that he will fail to press it.

This suggestion leaves out something important. Doesn't Sam's having complete control over which key he presses require his having complete control over which key he *tries* to press? What might that amount to? Here is a suggestion to consider. For Sam to have complete control at the time over which key he tries to press is for the following things to be true: regarding each key, when Sam hears the tone, he is able to decide to press it right then and, regarding each key, if he decides to press it right then, there is no chance that he will fail to try to press it.

There is a predictable worry about this suggestion too, of course. Doesn't Sam's having complete control in this scenario over which key he tries to press depend on his having complete control over which key he *decides* to press? What does that amount to?

When faced with my question about complete control over key presses, I looked to trying for an answer. And when faced with a parallel question about trying, I looked to deciding for an answer. Now that the question is what it is for Sam to have complete control over which key he decides to press, where should I turn?

Both of the suggestions I considered have a "no chance" clause. The reason for this is obvious. Suppose that the keyboard Sam is using has a randomizer on it that ensures that there is always a small chance that a key he is trying to press will stick and fail to make contact with the switch under it. (Recall the definition of a key press above.) Then Sam never has complete control over whether he presses the Q key or the P key. Suppose now that a randomizer has been installed in Sam's brain that ensures that there is always a small chance that his proximal decisions to press a key—his decisions to press a specific key straightaway—will not be followed by a corresponding attempt. Then Sam never has complete control over whether he tries to press the Q key or tries to press the P key (at least when no alternative route to attempt-production is in use—that is, no route that does not include decisions).

These observations prompt the following two questions. Might Sam nevertheless freely have pressed the Q key the last time he pressed it? Does a satisfactory account of a person's having complete control over whether he decides to A or decides to B have a "no chance" clause?

The correct answer to the first question, I believe, is yes, provided that free actions are common in Sam's world and agents can act freely when their options are of the kind featured in Buridan's ass scenarios. To see why, compare Sam's pressing the Q key with Al's voting for Gore or Ann's scoring the goal.

What about the second question? According to a standard libertarian view, if a person's deciding to A is to be a directly free action, there was a chance, right up to the time at which the decision was made, that he would not decide then to A. But this alone does not obviously commit a proponent of this view to claiming that no satisfactory account of a person's having complete control over whether he decides to A or decides to B can have a "no chance" clause. One reason is that the option of claiming that having complete control over whether one decides to A or decides to B is *not required* for directly freely deciding to A is not obviously a nonstarter.

Assessment of this option would benefit from an acceptable account of what it is to have complete control over whether one decides to A or decides to B. It can be said that having the kind of control at issue is a matter of its being *entirely up to* the agent whether she decides to A or decides to B or a matter of the agent's having the power to *determine* whether she decides (or intends) to A or decides (or intends) to B. But it is not as though the key terms here carry their meanings on their faces. Just as I would like to be told what it is to have complete control over whether one decides to A or decides to B, I would like to be told what it is for it to be entirely up to an agent whether she decides to A or decides to B and what it is to have the determining power at issue. From my point of view, it would be wonderful if the initial answers to my questions did not mention agent causation and then someone explained why agent causation is supposed to be needed for complete satisfaction of the conditions identified in those answers. But that wonderful state of affairs might not be in the cards. Owing to my ignorance about how I am supposed to understand the key expressions (for example, "complete control over"—or its being "entirely up to one"—whether one decides to A or decides to B), I do not know whether the (alleged) phenomena they are supposed to pick out can be given informative preliminary characterizations that do not appeal to agent causation.

Return to Sam. Imagine now that one thing that can happen when he detects the "decide" signal is that his ability to make decisions is temporarily eliminated. Sam is paid one dollar for each button press; so the temporary loss of the ability has a down side. When the signal is emitted, the following things are true, by hypothesis: Sam is able to respond to it with a decision to press Q; Sam is able to respond to it with a decision to press P; there is a chance that Sam will very soon temporarily lose his ability to make decisions. (Bear in mind that it takes a few milliseconds to detect the signal.) Consider a trio of possible worlds in which everything is the same up to the time the "decide" signal is emitted. In *W1*, Sam proceeds to decide to press Q; he makes this decision at *t*. In *W2*, he decides at *t* to press P. In *W3*, at *t*, his decision-making ability is temporarily eliminated. When he detected the signal, did Sam have complete control over whether he would decide to *A* or decide to *B*?

Compare *W1* with *W3*. The difference between them at *t* would seem to be just a matter of luck, and the same goes for the difference at *t* between *W2* and *W3*. But if these differences are just a matter of luck, are not Sam's deciding to press *Q* and his deciding to press *P* partly a matter of luck? If Sam had had the bad luck at *t* in those worlds that he had at *t* in *W3*, he would not have made either decision. Someone may assert that even if these decisions are partly a matter of luck, Sam had complete control in *W1* and *W2* over whether he decided to press P or decided to press Q, so long as he exercised agent-causal power when he made his decision. Because I do not know how "complete control" over this is supposed to be understood, I do not reject the assertion. In any case, perhaps partly because I am still in the dark about what the complete control at issue is supposed to be, it makes sense for me to continue to take comfort in *LUCK* and continue to offer my own solution to the problem of present luck (Mele 2006, chap. 5).

My solution—which makes no appeal to agent causation—has the virtue of being fairly easy to understand. In that respect, at least, it has an advantage over proposed agent-causal solutions to the problem of present luck that leave us in the dark about how such key terms as "complete control" and "determine the decision" are to be understood.

Acknowledgments

I am grateful to Meghan Griffith for comments on a draft of this essay.

References

Clarke, Randolph. 2003. *Libertarian Accounts of Free Will*. Oxford: Oxford University Press.

———. 2005. "Agent Causation and the Problem of Luck." *Pacific Philosophical Quarterly* 86:408–21.

Frankfurt, Harry. 1988. *The Importance of What We Care About*. Cambridge: Cambridge University Press.

Griffith, Meghan. 2010. "Why Agent-Caused Actions Are Not Lucky." *American Philosophical Quarterly* 47:43–56.

Kane, Robert. 1996. *The Significance of Free Will.* New York: Oxford University Press.

———. 1999. "Responsibility, Luck, and Chance: Reflections on Free Will and Indeterminism." *Journal of Philosophy* 96:217–40.

Kaufman, Arnold. 1966. "Practical Decision." *Mind* 75:25–44.

McCann, Hugh. 1986. "Intrinsic Intentionality." *Theory and Decision* 20:247–73.

Mele, Alfred. 1992. *Springs of Action.* New York: Oxford University Press.

———. 2003. *Motivation and Agency.* Oxford: Oxford University Press.

———. 2005. "Libertarianism, Luck, and Control." *Pacific Philosophical Quarterly* 86:395–421.

———. 2006. *Free Will and Luck.* New York: Oxford University Press.

———. 2010. "Moral Responsibility for Actions: Epistemic and Freedom Conditions." *Philosophical Explorations* 13:101–11.

———. 2013. "Moral Responsibility and the Continuation Problem." *Philosophical Studies* 162:237–55.

O'Connor, Timothy. 2000. *Persons and Causes.* New York: Oxford University Press.

———. 2011. "Agent-Causal Theories of Freedom." In *Oxford Handbook on Free Will*, ed. Robert Kane, 2nd ed., 309–28. New York: Oxford University Press.

Pereboom, Derk. 2001. *Living Without Free Will.* Cambridge: Cambridge University Press.

Pink, Thomas. 1996. *The Psychology of Freedom.* Cambridge: Cambridge University Press.

Searle, John. 2001. *Rationality in Action.* Cambridge, Mass.: MIT Press.

Taylor, Richard. 1966. *Action and Purpose.* Englewood Cliffs, N.J.: Prentice-Hall.

CHAPTER 6

YOU MAKE YOUR OWN LUCK

RACHEL MCKINNON

1. Luck Properly Under Control

This essay is composed of two principal projects. The first project is to discuss some popular sayings about luck, and to make sense of them in light of recent theories on the metaphysics of luck. The three that I will focus on are: "You make your own luck," "You have to be good to be lucky," and "Luck had nothing to do with it." The second project is to further develop my view on the metaphysics of luck given in "Getting Luck Properly Under Control" (McKinnon 2013), and to apply it more explicitly to a relatively well-established distinction between two kinds of luck—environmental and intervening luck—and how they impact attributions of credit. Specifically, I plan to draw out a deeper understanding of the relationship, and differences, between luck, skill, credit, and achievement.

I should briefly note what this essay is *not* about, though: it's not specifically about what view we should take on how luck undermines *knowledge*. What I have to say about the nature of luck certainly bears on discussions of knowledge as creditworthy belief, but I do not discuss knowledge specifically in any detail. Moreover, when I speak of credit and creditworthiness of actions (and beliefs, when we're talking about knowledge) I don't mean praiseworthiness of actions: I'm speaking of whether we credit an outcome to an agent's action. That is, I mean to speak of how we attribute outcomes to an agent's action.

The Philosophy of Luck, First Edition. Edited by Duncan Pritchard and Lee John Whittington.
Chapters © 2014 Metaphilosophy LLC and John Wiley & Sons Ltd, except for "Luck as Risk and the Lack of Control Account of Luck" © 2015 Metaphilosophy LLC and John Wiley & Sons Ltd. Book compilation © 2015 Metaphilosophy LLC and John Wiley & Sons Ltd.
Published 2015 by John Wiley & Sons Ltd.

In the next section I focus on the three locutions listed above. But before turning to them, I want to recapitulate the view of the metaphysics of luck I put forward in McKinnon 2013.

> I make the case that we can better understand the nature of luck by drawing an analogy with the expected value of wagers. A long series of wagers, each with a determinate expected value, itself has a determinate expected value. It's unlikely, however, that the outcome of the series will be the expected value. I argue that the difference between the actual results and the expected value is what we call luck. When we consider how actions, specifically when exercising a skill or ability, are similarly probabilistic, we can call the expected results of a series of actions "skill." Then, mutatis mutandis, the difference between the expected outcomes and the actual outcomes of a series of actions is what we call luck. Moreover, like a single wager viewed as part of a series of wagers, I propose that we view the outcome of an individual action as one element of a larger series of trials. Its status as creditable or lucky depends essentially on its place within that series. I subsequently argue that agents deserve credit for an outcome proportional only to their skill. Insofar as an agent's obtaining an outcome involves good luck, we should remove credit proportional to the good luck; similarly, insofar as an agent's obtaining an outcome involves bad luck, we should attribute credit proportional to the bad luck. (2013, 497)

Call this the Expected Outcome View of luck (EOV). The central example is someone placing a bet in poker. In many cases, we can precisely calculate the expected value of a wager. For example, if we're playing Texas Hold'Em and you go all-in pre-flop, then if we know your cards and we know my cards, we can precisely calculate the expected value of my call. If, for example, you simply bet $1,000 into an empty pot, and I have to either call $1,000 or fold, and I hold pocket aces (to your pocket twos), then I'm roughly 80 percent likely to win. The expected value (EV) of my bet is determined by the following (simplified) formula:

$$EV = (\text{Probability of Winning} \times \text{Amount Won})$$
$$- (\text{Probability of Losing} \times \text{Amount Lost})$$

In this case, my call with pocket aces has an expected value of +$600. EV can be positive, neutral, or negative. And this +$600 represents what I can expect to earn *on top of* getting my $1,000 bet back. This would be a *very* good bet for me to make, and assuming that I'm not extremely loss and risk averse (in which case I probably shouldn't be playing poker for these stakes), I should make this call.

One important point is that the result of this hand will be binary: either I win the $2,000 pot (a net gain of $1,000) or I lose my $1,000. Obtaining the +$600 EV isn't a possible outcome. So when I win, I win more than is "expected," as determined by the EV. So if I play only this one hand, on my view of luck, I'm lucky to the tune of the extra $400 above the EV of my bet. Relatedly, if I were to lose, I'm unlucky, and we can easily quantify that: I'm $1,600 short of what we expected me to have at the end of the bet (since I would be left without my $1,000 and without the +$600 EV of my bet).

To turn from betting to a more obviously physical action, take a professional basketball player such as Steve Nash. His skill in making free throw shots in game situations is such that he makes 90 percent of his shots. Making some, admittedly heavy, metaphysical assumptions, we can rephrase this to say that he's 90 percent likely to make any given free throw shot in a game. Now suppose that in a season he takes a hundred free throw shots and makes ninety-eight. On my view of luck, the expected outcome of this series of shots is for Steve to make ninety. Since he makes eight more than expected (assuming his skill is, for example, consistent throughout the season), he's lucky to the tune of eight shots. However, I argued in McKinnon 2013 that we can't specify *which* eight are attributable to luck. We can only know that eight are attributable to luck.[1]

The essence of my view is that we deserve credit for the expected value—or expected "outcome"—of our actions. If I can only win or lose, and the expected value is in between those options, then when I win, I don't deserve credit for everything that I've won; and when I lose, I deserve more credit than what I have (i.e., more than nothing).[2] The problem comes, I think, in that we often view luck and skill binaristically: either an action is attributable (entirely) to luck, or it's attributable (entirely) to skill. However, on my view, attributions of credit need to be more nuanced: we deserve credit for our skill proportional to the expected value (or expected outcome) of our actions. We should thus be more parsimonious in doling out credit when we're successful—i.e., when we come in above the expected outcomes—and more generous in doling out credit when we're unsuccessful—i.e., when we come in below the expected outcomes.

Finally, it's a critical feature of my view that part of what it means to have a particular skill is that one's actions in deploying that skill have a particular expected value or outcome. And if the skill is fallible—i.e., there's no guaranteed success when one exercises the skill—then sometimes, *even when*

---

[1] One might think that this is an odd way of putting things. If I buy a lottery ticket for $1 in a very large lottery, I may "expect" to lose the dollar. Suppose the expected value for my buying this ticket is $1. If I win the $1,000,000 first prize, one might say that I was lucky in winning the $1,000,000; one wouldn't be inclined to say that I was lucky only for the $999,999 difference between the outcome and the expected outcome. My argument in McKinnon 2013 is that this way of thinking is wrong.

[2] Here's a potential worry: one might think that on my view any personal best performance (e.g., squatting a particular weight or having my fastest 10 km run) would count as lucky. That's not necessarily the case. While it's true that some personal bests will include some luck (perhaps things "clicked" for me in a way that results in a personal best), if I have improved my ability, then the expected outcome of my performances will change. If I spend time doing interval training to increase my running speed, then it's hardly lucky that I improve my 10 km personal best.

one exercises the skill, one will fail to achieve one's goal. When Steve Nash takes a free throw, because his skill at free throws is fallible, sometimes, even when he's properly exercising his skill, he'll miss. It's not the case that he's exercising his skill only when he's successful, or that whenever he misses he's not exercising his skill (although there may be some misses that are due to his not exercising his skill). In fact, we expect him to miss some of the time (indeed, we expect him to miss 10 percent of the time). So failing to succeed in obtaining the goal(s) of one's actions doesn't mean that one failed to exercise one's skill.

2. Lucky Locutions

In this section I discuss three somewhat common locutions about luck: "You make your own luck," "You have to be good to be lucky," and "Luck had nothing to do with it." First, what does it mean to say that someone creates her own luck? At least colloquially speaking, luck is conceived as something out of an agent's control. So how could an agent increase or decrease the likelihood that she'll be lucky? I'll argue, however, that by applying my view of the metaphysics of luck, we can understand a sense in which an agent can create her own luck.

This sounds paradoxical probably because it's common to consider lucky events as those that are out of our control. I'm lucky in winning a lottery, it seems, partly because it's so unlikely that I'll win, but also because my winning isn't under my control. It's not like my making a difficult shot in pool. At least when I make a cross-table jump shot, it's *my* (relatively low) skill in doing so that causes the outcome. So in a pretheoretical view of luck, being out of our control seems like a plausible necessary condition.

In this vein, Daniel Statman offers the following definition of luck: "Good luck occurs when something good happens to an agent P, its occurrence being beyond P's control. Similarly, bad luck occurs when something bad happens to an agent P, its occurrence being beyond [P's] control" (1991,146). As I discuss in McKinnon 2013, there are a number of problems with this account of luck. It is neither a necessary nor a sufficient condition for some event's being lucky that it be out of our control. Andrew Latus (2000) gives an example where lack of control is not sufficient: although the sun's rising might, on some views, be out of my control, it's not particularly "lucky." And Jennifer Lackey (2008) offers examples where *that* an agent is in control is a matter of luck, such that the outcome is lucky but the agent is in control of the outcome. So there are a variety of problems with a control view of luck.

What is more important, though, is that the concept of "control" is fraught and underspecified.[3] What does it mean to be in control of a

---

[3] Coffman (2007, 2009), Riggs (2009), and Levy (2011) have all given attempts to elucidate the concept of control in the context of luck. I discuss problems with Riggs's account in McKinnon 2013.

situation or an outcome of an action? Steve Nash is, presumably, somewhat in control of whether his free throw will be successful, but his skill is fallible. His taking a shot, even while exercising his skill, only has a 90 percent probability of being successful. Moreover, even in cases where we have a strong intuition that we're in control of the outcome, often we forget or ignore many ways in which we're not "fully" in control. So for these, and other, reasons, we should jettison views that depend, strictly speaking, on a naïve view of control.

At this point I want to make a distinction between an *ability* and a *skill*. Too often in the literature the two have been used interchangeably.[4] Ability is sometimes used to refer to one's reliability in producing a particular outcome, such as Steve Nash's 90 percent likelihood of successfully making any given free throw shot. However, I think that this glosses over an important distinction between ability and skill. Subject S has ability A by performing action ϕ to produce outcome O in a range of contexts C iff were S to ϕ in C there would be a nonzero probability that S will O by ϕing in C.[5] I have the ability, for example, to make a half-court basketball shot (by moving my body a particular way), even though it's extremely unlikely that I will succeed in making the shot. Contrast ability, then, with skill. I have both the ability and the skill, for example, to make a basketball shot from $1''$ away from the basket by moving my body in a particular way. I have the ability but *not* the skill to make half-court basketball shots.

One can have the ability to O in C even when the probability that one will succeed is remote, but one does not thereby have the skill to O in C. What it means to have a skill is that one ϕs with sufficient regularity and in the right way(s). I'm a skilled competitive badminton player, for example. One thing I am skilled at is hitting a smash down the line (as opposed to, say, other skills I have, such as smashing toward the middle of the court or crosscourt). This requires moving my body in a particular, relatively small, range of ways to execute the shot. I can't move my body any which way, or it wouldn't constitute a smash, let alone one in the right direction (down the line, in this case). So having skill Q (and, to a lesser degree, having ability A) means that I ϕ in the right ways.

Furthermore, just because I have the skill, Q, to O by ϕing in C doesn't guarantee either that I will properly execute my skill or that my properly executing my skill will guarantee that I will O. In the first case, I might be distracted and so I don't ϕ in the right way required to constitute my skill. In order to smash down the line in badminton, I ought to contact the shuttle in front of my body. But if I'm momentarily distracted, say by a camera flash in the audience, the hesitation might mean that I contact the shuttle not in front of my body but behind me. Now suppose that when I execute my skill properly, I contact the shuttle in front of my body, and the smash goes where I want it to 95 percent of the time. However, when I'm distracted

[4] Notably, recently, by Greco (2012) and Turri (forthcoming), among many others.
[5] Here we assume that it's possible for S to ϕ.

and I don't contact the shuttle in front of me, it goes where I want it to only 5 percent of the time. My skill amounts, on my modal, probabilistic view of expected outcomes, to my smashes going where I want them to 95 percent of the time. So if I move my body in the relevant context, C, whereby I'm successful in having the shuttle go where I want it to (O) only 5 percent of the time, then this is not my exercising my skill (Q). I still *have* the skill, but I didn't exercise it in the context. At most, I exercised my unreliable ability (A) to hit a smash where I want to (O) when the shuttle is behind me.

In the second case, I exercise my skill, but I fail to obtain the desired outcome (O). This is a simple consequence of fallible skills. My skill in hitting a smash amounts to my hitting the shuttle where I want it to go 95 percent of the time. This means that *even when I'm exercising my skill*, I'll miss 5 percent of the time. Missing, or failing to obtain the desired outcome (O), doesn't entail that I didn't (properly) exercise my skill.[6] This is partly why the control view of luck is so fraught: if skills are fallible, then we're never "fully" in control of our actions obtaining their desired outcomes, however we cash out our concept of control.

But now that this detour is out of the way, I can return to discuss the first locution, "You make your own luck." On my view of luck, skill, and ability, it's possible that one can make one's own luck—in a sense, at least. One way for an agent to get lucky is for outcomes to happen more frequently, modally speaking, than we expect them to given an agent's skill in a particular range of contexts. Again suppose my "skill" at hitting badminton smashes down the line when the shuttle is behind me amounts to being able to hit the shuttle where I want to only 5 percent of the time. Suppose that I take twenty such shots during a match, and all twenty are hit down the line. I have vastly outperformed the expected outcome of *one* successfully hit smash down the line (expected outcome = 1/20 = 5 percent). Now certainly it's *possible* that I can hit all twenty down the line, but the likelihood is remote (it's $p(x) = 0.05\hat{\ }20$). And on my view of luck and skill, we can attribute one of my successes to skill, and the rest to luck (alternatively speaking, 5 percent of all of my outcomes are attributable to skill, and the rest to luck).

Crucially, though, if my skill in this range of contexts is to succeed 5 percent of the time, there's no way for me intentionally to increase the number of times I will get lucky. There is a way, however, that agents have some control over how likely they are to get lucky: by becoming more skillful. Now, someone with a high degree of skill (say, 95 percent likely to succeed in C)

---

[6] I think this has important implications for, for example, virtue epistemologists who take the view that knowledge is exercising a cognitive or epistemic skill. On the view I'm painting, epistemic justification amounts to properly exercising one's cognitive or epistemic skill, and knowledge is when one's properly exercising one's cognitive or epistemic skill results in believing something true. One consequence of my view is that some Gettier cases count as knowledge, whereas others do not.

won't be more likely to experience luck than someone with a relatively low degree of skill (say, 5 percent likely to succeed in C) per se. As people say in sports, "in the long run" good and bad luck even out, and all that's left are differences in skills between competitors. I think that's largely right. Due to the structural constraints of various games and competitions, however, better players tend to *stay in the game longer* and thus tend to have more opportunities for luck to rear its head.

Let's take a toy example. If I take one shot where I have a skill of 95 percent likelihood of succeeding, then I'm only 5 percent likely to be lucky in my performance. However, the probability of my being lucky at least once if I get to perform that same skill, in the same context, ten times is higher. The probability that I will get lucky at least once in these ten attempts goes up to 40 percent. So simply by getting more *opportunities* to get lucky, by being in the game longer, I will be more likely to get lucky.[7] Now while the more opportunities one has increases the likelihood of experiencing good luck at least once, this also increases the likelihood, mutatis mutandis, of experiencing bad luck at least once. The locution tends only to apply to experiencing good luck, though.

Moving back to a betting example, let's consider a poker tournament. The tournament starts with a thousand players, until it's down to two, and then down to the winner. Poker is a game where probabilities are more transparent to the competitors than in other activities like playing basketball. If two players go all in where one is 95 percent likely to win (and thus the other is 5 percent likely to win), the person making the bet with 95 percent probability will be more likely to last longer in the tournament. Now, it's common in poker tournaments for players who last deep into the tournament to get into lots of situations where they're only 50 percent likely to win. Those who last the longest tend to be those who don't lose any (or many) of these "coin flip" hands.

Now suppose that we compare two players and how lucky they are over the course of a tournament. Annie goes all in on her first hand, in a coin flip situation, and loses. Her outcome is below the expected value of her action, and so she's unlucky to that degree. Bonnie goes all in on her first hand, in a coin flip situation, and she wins. Symmetrically with Annie, mutatis mutandis, Bonnie is a little lucky. Now suppose that Bonnie is a *very* good poker player. She uses her skill to maneuver the tournament so that she only gets into coin flip situations when she's not the player all in, at risk of being eliminated from the tournament. Rather, we may suppose, she only gets into coin flip situations when the pot is at most 10 percent of her remaining chips. This means that even if she were to be "unlucky" in a particular coin flip hand, she won't be eliminated from the tournament. This results in her, we may suppose, getting into ten coin flip hands over the course of the

---

[7] This is a variation on the theme of "You can't get lucky if you don't play."

tournament. One of her skills involves *when to enter* into coin flip hands, and one consequence of her deploying this skill, in this case, is that she gets more chances to get lucky in hands.[8] Annie only had one opportunity, and she lost. Bonnie has had many opportunities, so the probability of her getting lucky in at least one of these hands has increased. The probability of getting lucky in one coin flip hand is 50 percent (half the time one is unlucky, the other half one is lucky). But the probability of getting lucky (i.e., winning) at least one hand out of ten is 99.9 percent. By getting more chances at such situations, Bonnie is more likely to get lucky than Annie.

My argument is thus that one thing that increased skill allows us to do is stay in the game longer, so that we have more opportunities to get lucky.[9] And in many cases, we have at least some control over our skill: we can improve through training and practicing, for example. This is no different in poker: one can "study" poker in order to improve.[10] And in this sense, but only this sense, I think we can make sense of how it can be true that "you create your own luck."[11] We can create our own luck (probabilistically, anyway) by increasing our skill, because increasing skill makes it more likely that we have more opportunities to get luck, which means that we're more likely to get lucky than those with lower skill.

Related to this line of argument is how my view offers an interpretation for the second locution: "You have to be good to be lucky." We find this locution, with some regularity, in sports commentaries.[12] Of course,

---

[8] This is related to Sosa's (2011) concept of meta-aptness: a competence in when to exercise a competence, such as a competence in when to (attempt to) competently shoot an arrow at a target.

[9] One might think that there's an alternate reading of this: skill allows us to win faster, which means that there are fewer opportunities to experience bad luck. It's certainly true that exercising a higher skill in, say, a badminton match means that I'll win more quickly and avoid some opportunities to suffer bad luck by, say, spraining an ankle. I don't think that this is an alternate reading: it's a consequence of my view. By winning more quickly, there are fewer opportunities to get lucky (in either the good or bad sense). And winning more quickly is a consequence of having higher skill. There's thus a common cause: the higher expected outcome means that the game ends more quickly (for the more skilled person), and *consequently* reduces the chances to suffer bad luck. My discussion in this section, however, is about the locution "You make your own luck." In some cases, higher skill means games or tournaments lasting longer, which increases the likelihood that one will experience luck. It doesn't follow from this that having higher skill *guarantees* that one stays in a game longer: it just makes it more likely, in some contexts. In others, it makes it more likely that the game will end more quickly (with the more skillful person tending to win).

[10] I should know: I played professionally for six years. This regularly involved playing for two to three hours each day, bookended by up to four hours of studying strategies, situations, and player tendencies.

[11] This phrase is used in the 2008 Batman movie, *The Dark Knight*. Rachel sees that TwoFace has marked both sides of his coin, so that there's no luck involved in his flips. She says, "You make your own luck." This utterance is infelicitous, if read literally: there's *no luck* to be made here. What TwoFace is doing is taking luck's role out of his coin flips: he can't lose.

[12] Here's but one quick example: http://www.raptorshq.com/2011/4/27/2135884/you-gotta-be-good-to-be-lucky-and-lucky-to-be-good (last accessed Dec. 17, 2013).

interpreted literally this is false: anyone with an ability can get lucky. That's analytically true, following the definition of ability given above. I think, however, that the real meaning behind the locution is very much in line with what I've just had to say about how more highly skilled people are, statistically speaking, more likely to get lucky than lower-skilled people.

Alexander Ovechkin is an amazing professional hockey player. In one particular goal, he broke for the net, fell, and from his back, above his head on the ice, he shot the puck and scored.[13] Given that he was sliding on his back in that position, very few players could have made that shot in game conditions. In fact, we may suppose that, modally speaking, most times Ovechkin himself is put in that situation, he would miss. However, and here's the important point, very few players are good enough to make the break for the net that he did: he had to be good enough to *put himself in a position* to get lucky (insofar as his goal is even partially attributable to luck). Being skillful opens up more opportunities for an agent to experience (good) luck. Of course, it doesn't follow on my view that a skillful agent *will* be more lucky than a less skillful agent, but the more skillful agent will be more likely to be lucky than the less skillful agent.

This brings us, finally, to the third locution about luck: "Luck had nothing to do with it." I think we tend to find this as a response from someone who successfully (in that the desired outcome was obtained) performed some action, and who has been challenged by someone else for that success being attributable mostly (if not entirely) to luck. For example, in March 2013 I went to a casino to play some $1/2 No Limit Texas Hold'Em poker. Given that I used to be a professional, my skill level for these stakes of play is very much higher than that of almost any other player I would likely encounter in the context. Accordingly, one thing that my increased skill allows me to do is to play a wider variety of (lower-quality) starting hands profitably. For example, against other good players, it's unlikely that I would call a pre-flop raise holding 7d4d (d = diamonds). But against bad players, this sort of hand can be profitable in the right situations.

In the hand in question, someone raised to $8, and three people called. Knowing that I would be "on the button," which is the most powerful position post-flop, and given that there would be at least three other people in the hand (and that a pre-flop re-raise was unlikely), my calling the $8 is a good play: it has positive expected value. As the hand played out, the flop gave me an inside straight draw: I needed a 5 to make a straight (3–7). Bad players tend to bet too small, which is a mistake that good players can exploit. So I got to call the flop and turn bets cheaply to see the river; each call also had a positive expected value, particularly given that it was likely that I could bet or raise the river and get called if I hit my straight (we call

---

[13] Here's a clip of the goal: http://www.youtube.com/watch?v=LTBUTy8QKgM (last accessed Dec. 17, 2013).

this "implied odds"). As "luck" would have it, the river was the 5, I bet big, and got called.

The point of this vignette is to showcase that each decision had a positive expected value, that I knew this, and that my play was (thus) skillful. But this was opaque to the other players, partly due to their lower skill. They couldn't see how calling as I did pre-flop with such a "poor" hand could possibly be a good play. So someone said, "You just got lucky." An option for reply on my part could have been, "Luck had nothing to do with it."[14] Here, I'd be responding to the interlocutor's *under-attribution* of skill to my performance. My response indicates this, and is meant to suggest that while my performance may have appeared to have been the result (largely) of luck, it was really due to skill.[15] But importantly, I think, even if I were to respond this way, I wouldn't literally mean that luck had *nothing* to do with my success. I was lucky in a number of ways: that the river was the 5 that I needed was, in a sense, lucky—although my calls having a positive expected value is true irrespective of the river being a 5—as was the fact that I didn't have a heart attack moments before the river was a 5. To recapitulate, then, "Luck had nothing to do with it" is a response to an interlocutor's under-attribution of skill; it's not literally expressing the claim that luck had *nothing* to do with the outcome, since that would be false.[16]

In this section, then, I've shown how my account of the metaphysics of luck, and how it lays out how we ought to go about making attributions of skill and credit, explicates three potentially puzzling locutions: "You make your own luck," "You have to be good to be lucky," and "Luck had nothing to do with it." For the first locution, it's possible for agents to create their own luck by increasing their skill, because being more skillful tends to lead to more opportunities to be lucky, which makes it more likely for agents to be lucky. For the second locution, another consequence of being skillful is that it tends to create more opportunities to be lucky than being less skillful does. So while the locution, literally interpreted, is false, it has a nonliteral meaning that expresses the feature that people who are more skilled tend to have more opportunities to benefit from luck. Finally, when it comes to

---

[14] Of course, one ought never to discuss and explain strategy in a situation like this, so I didn't respond with "Luck had nothing to do with it." In other contexts, I would have.

[15] Of course, indicating that I might be skilled is the last thing I would want to do in this context, which is why I didn't respond this way when challenged. Rather, I agreed that, yes, I had gotten extremely lucky (because I want other players to under-attribute skill to me).

[16] There might be some cases where it's literally true that luck had nothing to do with an outcome. However, I find it hard to conceive of cases. Say I'm playing a strategy game and I've anticipated a particular gambit by my opponent. Upon learning that I've anticipated her gambit, she might say, "You got lucky that I did that!" I might respond with, "Luck had nothing to do with it. I know how you play." But this isn't literally true: the expected outcome of our game wasn't that, certainly, she would play the gambit. There was still *some* luck involved: she may have unpredictably radically changed her strategy in a way that I couldn't anticipate.

the third locution, while it is also literally false, we tend to see it used as a response to an interlocutor's under-attribution of skill toward a particular outcome of an action.

3. "Intervening" and "Environmental" Luck

In the remainder of this essay I take up some recent distinctions between different kinds of luck and how they affect attributions of credit. I argue that what has become a relatively well-established distinction—between "intervening" and "environmental" luck—is actually quite dubious. It's important to note at the outset that one need not adopt a particular view on the metaphysics of luck, let alone my own EOV, to see that there's something amiss with the distinction between intervening and environmental luck. However, I argue at the end that combining a counterfactual view of causation with my expected outcome view of luck helps us better understand how to attribute credit in cases previously discussed in terms of intervening and environmental luck.

Duncan Pritchard argues for the following two necessary (but not jointly sufficient) conditions for luck:

L1:   If an event is lucky, then it is an event that occurs in the actual world but which does not occur in a wide class of the nearest possible worlds where the relevant initial conditions for that event are the same as in the actual world.

L2:   If an event is lucky, then it is an event that is significant to the agent concerned (or would be significant, were the agent to be availed of the relevant facts). (Pritchard 2005, 125, 132)[17]

While I argue against this as an adequate view of the metaphysics of luck in McKinnon 2013, my focus here is different. Starting with Mylan Engel (1992), and further developed by Pritchard (2005, 2007, and elsewhere), philosophers have distinguished between *veritic* and *evidential* luck. In the epistemic domain, these concern how a true belief is formed. I'll start with the latter. A belief is evidentially luckily true when the agent's coming to truly believe something is the result of a relevant (cognitive or epistemic) skill or ability, but *that* the agent is in a position to know is a matter of luck.[18] Standing where I am, I don't know how many coins are under my desk. If, however, I were to "luckily" trip and fall such that I can see under my desk, and the lighting and my eyesight are sufficiently good for me to

---

[17] Although he seems to offer them as jointly sufficient, Riggs notes that Pritchard has communicated that L1 and L2 are not offered as jointly sufficient. See Riggs 2009, 207, n. 3.

[18] So far as I know, no one has put this in terms of "epistemic" skill or ability. There are mounting objections to whether knowledge is best understood in terms of "cognitive" skill. For example, see Miracchi (forthcoming). I suspect that shifting to discuss "epistemic" skills and abilities might avoid these concerns, but that's a matter for a different project.

reliably see that there are three coins under my desk, I can come to know that there are three coins under my desk. My belief was formed through a relevant cognitive ability (vision, for example), but that I was in a position to deploy my ability was a matter of luck. According to Pritchard (2005, 2012, and elsewhere) and many other anti-luck epistemologists, evidential luck does not undermine knowledge.

A belief is vertically luckily true when the belief's being true is a matter of luck. There are different ways to cash this out, but those putting forward some version of anti-luck virtue epistemology tend to say that a true belief is veritically lucky when the agent's belief in something true doesn't *manifest* her (cognitive or epistemic) ability. I have at least two ways of truly believing what happens to be the weather in a different city: I can guess, or I can go online and look up the weather report. If I look up the weather report online, then my true belief that the weather in London, Ontario, is "light snow" will manifest—it will be appropriately "because of"—my ability to find a reliable source of weather information for London, Ontario. Consequently, I know what the weather in London, Ontario, is right now. However, I could also just guess that the weather in London, Ontario, right now is "light snow." If, however, my belief turns out to be true, then that I happen to form a true belief about the weather is a matter of luck, and my true belief doesn't manifest a relevant ability. Simply guessing just isn't a sufficiently reliable belief-forming process to confer epistemic justification on my belief. So even though I believe something true, since that belief was formed through guessing, I don't know what the weather in London, Ontario, is right now: veritic luck undermines knowledge. This is a central tenet of what has become known as anti-luck epistemology.

One central task of anti-luck epistemology is to put forward a theory of knowledge that precludes knowledge in cases of veritic luck. Another task is to determine how, exactly, veritic luck undermines knowledge. My intent in this essay is to take up the latter project. However, I focus almost exclusively on the *nonepistemic* examples Pritchard uses to make his case. Pritchard (2005, 2012) distinguishes between two kinds of veritic luck: *intervening* and *environmental* (see also Carter 2011 and Jarvis 2013). Moving from the epistemic domain to the sports performance domain, let's use a favorite example of virtue epistemologists: an archer shooting an arrow.[19] Gina is an expert competitive archer. She rarely misses the bull's-eye when she's aiming for it. She's also an expert at judging the wind and waiting for just the right time to take her shot.[20] Suppose Gina judges the wind well, and takes her shot. The arrow is headed directly for the bull's-eye, but moments before it hits the target, an unexpected gust of wind blows her arrow off target. In addition, though, a moment after this gust of wind

---

[19] The example is taken from Sosa 2007. See also Sosa 2011.
[20] In Sosa's (2011) terms, she's both apt and meta-apt.

blows the arrow off target, another gust of wind, coming from the opposite direction, blows the arrow back on target. Consequently, Gina hits the bull's-eye.

On Pritchard's view, this is a case of intervening luck: that Gina's arrow successfully hit the bull's-eye was a matter of luck, even though she exercised her archery skill. Consequently, she doesn't deserve (full) credit for her success. That is, we don't attribute credit for the success to her skill. Now, Gina exercised her archery skill and her skill at choosing when to take her shot. The problem is that the arrow's hitting the target doesn't sufficiently *manifest* her archery skills; rather, it's due to the canceling gusts of wind. Likewise, in the case of some Gettier cases in epistemology, even though a belief may be formed through a reliable process, and the belief is true, if there is intervening luck, the luck undermines an agent's knowledge.

Contrast intervening luck with environmental luck. Jane is submerged in water in a watertight glass box with a piano. She's an expert piano player and is tasked with playing a particular concerto that she knows well. The box is unsafe, though, and could give way at any moment. If it does, water will rush in, ruining her performance. Luckily for Jane (and her audience), the box holds, and she is able to successfully finish the concerto. Turning to traditional epistemology cases, when Barney is driving through fake barn county and happens to look at the one real barn among hundreds of barn facades (which from his viewpoint are indistinguishable from real barns), his true belief that the building he is looking at is a barn is a case of environmental luck. Here, intuitions differ on whether Jane deserves credit for her success, and on whether Barney knows that what he's looking at is a barn. Pritchard (2012) argues that they do not: environmental luck undermines credit, and thus knowledge.[21] Ernest Sosa (2007, 2011) and John Turri (2011) argue that agents do deserve credit for success in cases of environmental luck. But I want to set these issues aside.

For my part, I think that the distinction between environmental and intervening luck is not a clear one. And I think understanding why can help us resolve some of the dispute between the different intuitions people have on the matter. Moreover, by utilizing a counterfactual theory of causation, I think the expected outcome view of luck I've articulated in McKinnon 2013 and in this essay can help better explain what's going on. As I noted

---

[21] It's important, however, to note that Pritchard argues that knowledge-excluding environmental luck is compatible with cognitive achievements. Certainly, agents who truly believe in environmental luck cases believe the truth (that's an achievement). But since the achievement doesn't properly manifest the agent's cognitive skill, the agent fails to know. I think, however, that this is wrong, but discussing it in more detail would be beyond the scope of this essay. Gina deserves credit (in proportion to the expected outcome of her action) for her success in the environmental luck case, just as the pianist does for completing her concerto in the submerged box. Similarly, in the epistemic case, an agent in fake barn county deserves credit for her true belief: she *knows* even though there's environmental luck.

in the beginning of this section, however, I think there are problems with the distinction regardless what view of luck one adopts.

One problem I see with the distinction between intervening and environmental (veritic) luck is that what it means for something to "intervene" is underspecified. Consider two parallel archery cases. Elaborating on the case above, Gina is in an archery competition indoors. The hall has two sets of windows on the side. These windows happen to be open at the moment, and a button electronically controls the window shutters by the door of the hall. Gina takes her shot, and her shot is heading directly for the bull's-eye. The two unexpected, but canceling, gusts of wind blow into the hall and affect her arrow, and she hits the bull's-eye, winning the competition. This is a clear case of intervening veritic luck, in Pritchard's sense. On his view, Gina's successfully hitting the bull's-eye doesn't appropriately manifest her archery ability, because the two gusts of wind intervened, even though they canceled each other out.

Now take the same shooting conditions. Gina* takes the same shot, under the same conditions, with the same unknown gusts of wind on their way to the archery hall. However, just as the gusts are about to reach the windows, a bystander, James, accidentally hits the button activating the window shutters. The shutters close just before the wind gusts arrive at the hall, thus blocking the wind. Gina*'s shot, which was heading directly for the bull's-eye, hits the bull's-eye, and she wins the competition. On Pritchard's view, this would be a case of environmental luck and *not* a case of intervening luck. It would be like the submerged concert pianist, Jane: the water could have affected her performance had the glass enclosure broken, but it didn't; here, the wind could have affected Gina*'s shot had the shutters not been open, but it didn't. Here's the problem: don't the shutters "intervene" by blocking what would have been an intervening cause had the shutters not closed? It's not clear to me why we should distinguish between these two cases, Gina and Gina*, in terms of our evaluation of the creditworthiness of their performances; moreover, it's not clear that there wasn't any "intervening" in the Gina* case.

One worry I have is that Pritchard's distinction depends on something like, what I like to call, a "Don't touch my stuff" view of intervening causes.[22] Something (an agent, an object, etc.), X, intervenes with something else, Y, just in case X touches or comes into contact with Y. The wind intervenes in Gina's shot because it actually touches and physically moves the arrow (and the other gust moves it to the arrow's original flight path). But the wind doesn't intervene in Gina*'s shot because it is blocked by the window shutters and thus doesn't physically touch or move the arrow. I think this is an odd theory of causation and intervening. In short, it seems to presuppose that omissions and preemptions are not causes, which is

[22] My thanks to Nicole Wyatt for the phrasing.

highly controversial. To see why, it will be useful to consider what have become known as counterfactual views of causation.[23]

David Lewis (1973) offers one counterfactual analysis of causation: if *e* and *c* are both events that occur, then *c* causes *e* iff were *c* not to happen, *e* wouldn't have happened. Had Gina not taken the shot, then she wouldn't have hit the bull's-eye; therefore, her taking the shot caused (in part) her hitting the bull's-eye. Counterfactual views allow omissions and preemptions to count as causes, too. L. A. Paul and Ned Hall (2013, 195) offer one formulation: "Had an event type C occurred, event E would not have occurred." Take a simple case from Paul and Hall (2013, 180; see also McDermott 1995 and Collins 2000). Suzy throws a rock at a window. Billy is in position to block the rock (thus preventing the window from breaking). If Billy blocks the rock, he will "cause" the window not to break. Suppose, however, that he chooses not to block Suzy's throw, and the window breaks. On many views of causation, including counterfactual views, Billy's decision *not* to block Suzy's throw is one cause of the window breaking (in addition, for example, to Suzy's throwing the rock). It's true that had Billy blocked the rock (C), the window would not have broken (E). He didn't block the rock, and the window broke. While Billy didn't "act," his omission is still a cause of the window breaking.

We can easily connect this case to luck, too. Suppose that Billy has dared Suzy to throw a rock at the window. He has promised, repeatedly, to block the rock so that Suzy won't actually break the window. He even tells her that if she happens to break the window, he'll pay for it and take full responsibility. Billy just wants to test whether Suzy has the resolve to accept and act on the dare. Suzy throws the rock at the window, and Billy decides to be a jerk and doesn't block the rock. The window thus breaks. I think it's fair to say that Suzy was *unlucky* in breaking the window: she expected Billy to block the rock.[24] Not only would she not have thrown the rock without his taunting, she wouldn't have thrown it without his assurance to block the rock from breaking the window. My point here is that omissions, as causes, can also contribute to whether outcomes are lucky (or unlucky).

On this sort of counterfactual view of causation, Gina's shot, the wind, the shutters closing (or not), and James's pressing the button (or not)

---

[23] Certainly, the jury is still out on which theory of causation seems best. However, counterfactual views seem to have a number of advantages over a number of rivals. For a detailed treatment of the debates, see Paul and Hall 2013.

[24] I grant that some might not share this intuition. But this isn't about intuitions: treating Suzy's breaking the window as unlucky for Suzy is a consequence of my view (it's not part of my argument for it). The point is that the expected outcome, given Billy's genuine promise, is that he will stop the rock. When he doesn't, the actual outcome deviates from the expected outcome, which results in luck (either good or bad, depending on the direction of deviation). We could easily change the case such that Billy genuinely intended to stop the rock but suffers a fatal heart attack in the moment between Suzy's throwing the rock and his attempt to block it.

are all causes of Gina's hitting the bull's-eye (or not). There's no distinction between "intervening" (as physical touching) and "environmental" causes.[25] My point in raising the counterfactual view of causation is
twofold: first, omissions and preemptions are causes—or, if they're not,
those who wish to maintain the distinction between intervening and environmental causes, such as Pritchard, owe us a story about what view of
causation they're committing themselves to; second, combining a counterfactual view of causation with my expected outcome view of skill and luck
gives us different responses to key cases in the luck literature.

In this vein, I think we should treat the Gina and Gina* cases the same
with respect to giving Gina and Gina* credit for successfully hitting the
bull's-eye, and the success manifesting their ability. In Gina's case, had the
wind not happened, she would have hit the bull's-eye all the same. Given
that the two intervening winds cancel each other out, they have no net effect
on her shot. I think adopting a counterfactual view of causation helps us
understand why. First, just because the two canceling winds are causes in
Gina's arrow hitting the target, we can't forget that Gina is *also* a cause
of the arrow hitting the bull's-eye. Her part in the causal chain can't be
dismissed. Had she not taken the shot, the arrow wouldn't have hit the
bull's-eye, gusts of wind or not. Second, when taken as a set, had the two
winds not come into the archery hall, Gina's shot would have had the same
outcome: hitting the bull's-eye. This is why we should treat the assessment
of Gina's and Gina*'s performances equally. Other than the winds physically touching Gina's arrow (because of the omission of the shutters) but
not Gina*'s arrow (because they were blocked by the shutters), there's no
difference between the two performances and the two outcomes. In both
cases Gina and Gina* are integral causes of their shots hitting the bull's-
eye. While there are other factors than merely their skill in achieving the
outcome, they don't affect the expected outcomes of their shots. Both Gina
and Gina* executed their skill, and their outcomes *manifest* their skill. And
the probability that they would hit the bull's-eye, given their properly executing their skill, isn't affected by the presence or absence of the winds. That
is, the expected outcomes of both cases are identical.

---

[25] One might raise a worry about applying this to the epistemic case. The idea is that environmental luck may preclude an agent's knowledge merely because of the way the environment
is "unfriendly" to modally forming true beliefs in most nearby possible worlds (although one
forms a true belief in the actual world). Think of Ginet's (1976) barn facade county: Fred
happens to look at the one real barn, forms the belief "That's a barn," and in most nearby
possible worlds in which he looks at a barn-like object, he'll be looking at a barn, and his
belief will be false. The worry is that environments aren't events, so my objection, relying on
Lewis's counterfactual view of causes, is problematic. This might be true. I think it's irrelevant,
though: my purpose in this discussion is not to show how Pritchard's view of why environmental luck undermines knowledge is wrong (I take no position on that); rather, it's to show that
his distinction between environmental and intervening luck has problems, particularly in the
nonepistemic examples he uses.

In addition, I worry that treating the two cases differently glosses over just how causally messy our performances, successful and not, really are. Even in Gina*'s case, it's false that the air in the hall is perfectly still or even perfectly uniform, if there's a draft. When Sosa (2011) talks about an archer being skilled at choosing when to shoot, we might think of how Gina, especially if she's shooting in conditions of wind, needs to judge the wind in both choosing when to shoot and how to adjust her aim to compensate for the wind. But we often forget that shooting an arrow in the real world isn't like a video game (like archery in *Wii Sports*): the wind subtly fluctuates in unpredictable ways. Archers make their best guess that, over the long run, tends to have the good and bad luck even out, so that what's left is a relatively accurate measure of their skill.

The wind, however, is inherently nonuniform and unpredictable. Moreover, subtle undetected differences in an arrow (its weighting, the condition of the fletchings, and so on) affect the arrow's flight path. And the flight path itself is not a perfect parabola from the bow to the target: the arrow will spin and have subtle (or not so subtle) dips, lifts, and twists as it flies. An even more exaggerated version of this is professional darts: if one watches the flight path of a dart (even one that purposefully hits the bull's-eye), one might marvel at how anyone ever hits her intended target, as the arrow wobbles wildly in its flight path. And given that our performances are "messy" themselves—essentially, no two golf swings, by the same person, will be the same; no two dart throws are exactly the same—this often leaves our performances sensitive to minute changes in environment such as wind, air pressure, humidity, and so on. And yet the presence of all these unpredictable factors—ones that would "touch" our arrows, for example, and thus count as intervening luck in Pritchard's sense—don't incline us strongly to fail to treat the success as creditable to an agent's skillful performance. These factors are simply features of the context of performance, and they're very often chaotic and unpredictable.

Simply put, luck, even the supposedly problematic but underdefined "intervening" kind, often permeates our performances in unappreciated ways. A competitive ski jumper's coach stands at the bottom of the ramp judging the best wind conditions—within the jumper's time window for beginning her jump—but by the time the jumper has started her jump, the wind conditions will have changed in unpredictable ways. And these unpredictable changes have important effects on a jumper's distance and, thus, performance. Over a career, however, the small differences between the wind conditions from jumper to jumper, and jump to jump, tend to even out so that the better jumpers have better results. So on a day-to-day scale, various forms of luck permeate our performances. The hope is that in the long run these small fluctuations will even out, although they may not. But if the mere presence of this sort of luck undermines an agent's credit-worthiness for a successful performance, then I think we're in trouble: we wouldn't deserve credit for almost all of our successes. That is, our successes wouldn't be creditable to our (skillful) performances.

Fortunately, we aren't forced to view the presence of "undermining" luck as an all-or-nothing affair. EOV gives us recourse to a more finegrained analysis of how luck sometimes undermines creditworthiness, but only *partially.* It also gives us a way, partly, to quantify how much luck undermines the creditworthiness of a particular performance (viz., its outcome).

4. Moving Forward

Combining some version of a counterfactual view of causes with my expected outcome view of luck allows us to avoid a number of the problems plaguing other extant proposals, including Pritchard's modal account of luck. We need not make a distinction between different kinds of luck, intervening or environmental. It's simply not clear whether there's a clear distinction to be made. In attributing credit for a success to an agent's performance, we determine the extent of an agent's skill and the skill's expected outcomes for the context. To reiterate: S has skill Q to produce outcome O by ϕing in a range of contexts C iff were S to ϕ in C, S would O with sufficient regularity by ϕing in C. Q will have a determinate expected outcome in C.

Gina, for example, might be 99 percent likely to hit the bull's-eye, when she aims for it, when she exercises her skill in the context of the competition (given the environmental conditions, and so on). If, somehow, various environmental conditions conspire to affect her shot, in terms of causes, counterfactually understood, then we determine whether these additional causes affect the expected outcomes of her exercising her skill. Some such conditions will make her shots more likely or less likely to result in success. If these extra-skill causes make it more likely that Gina will succeed (perhaps someone has placed a powerful electromagnet in the bull's-eye that will essentially guarantee that she will hit the bull's-eye), then we reduce the amount of credit for the success attributable to her performance. If these extra-skill causes make it less likely, mutatis mutandis, we increase the amount of credit for the success attributable to her performance (as I argued in McKinnon 2013). If, however, these extra-skill causes have no net effect on the expected outcomes of the agent exercising her skill, then this has *no effect* on our attribution of credit for the success of her performance. This is in stark contrast to other views, such as Pritchard's, that suggest the agent deserves no (or little) credit for many of these performances. I think that that's the wrong result. And combining my expected outcomes view of luck with a counterfactual view of causation helps explain why.

Acknowledgments

My thanks to Clayton Littlejohn, Lisa Mirrachi, Matthew Benton, Liam Bright, Nicole Wyatt, Laurie Paul, Gina Angelea, Duncan Pritchard, Lee

Whittington, and the audience members at the Department of Philosophy Speaker's Series at the University of Calgary. This research was supported by funding from the Social Sciences and Humanities Research Council of Canada.

References

Carter, J. Adam. 2011. "A Problem for Pritchard's Anti-Luck Virtue Epistemology." *Erkenntnis* 78, no. 2:253–75.

Coffman, E. J. 2007. "Thinking About Luck." *Synthese* 158, no. 3:385–98.

———. 2009. "Does Luck Exclude Control?" *Australasian Journal of Philosophy* 87, no. 3:499–504.

Collins, John. 2000. "Preemptive Prevention." *Journal of Philosophy* 97, no. 4:223–34.

Engel, Mylan, Jr. 1992. "Is Epistemic Luck Compatible with Knowledge?" *Southern Journal of Philosophy* 30, no. 2:59–75.

Greco, John. 2012. "A (Different) Virtue Epistemology." *Philosophy and Phenomenological Research* 85, no. 1:1–26.

Jarvis, Benjamin. 2013. "Knowledge, Cognitive Achievement, and Environmental Luck." *Pacific Philosophical Quarterly* 94:529–51.

Lackey, Jennifer. 2008. "What Luck Is Not." *Australasian Journal of Philosophy* 86, no. 2:255–67.

Latus, Andrew. 2000. "Moral and Epistemic Luck." *Journal of Philosophical Research* 25:149–72.

Levy, Neil. 2011. *Hard Luck: How Luck Undermines Free Will and Moral Responsibility*. Oxford: Oxford University Press.

Lewis, David. 1973. "Causation." *Journal of Philosophy* 70:556–67.

McDermott, Michael. 1995. "Redundant Causation." *British Journal for the Philosophy of Science* 46, no. 4:523–44.

McKinnon, Rachel. 2013. "Getting Luck Properly Under Control." *Metaphilosophy* 44, no. 4:496–511.

Miracchi, L. Forthcoming. "Competence To Know." *Philosophical Studies*.

Paul, L. A., and Ned Hall. 2013. *Causation: A User's Guide*. Oxford: Oxford University Press.

Pritchard, Duncan. 2005. *Epistemic Luck*. Oxford: Oxford University Press.

———. 2007. "Anti-luck Epistemology." *Synthese* 158, no. 3:277–97.

———. 2012. "Anti-Luck Virtue Epistemology." *Journal of Philosophy* 109:247–79.

Riggs, Wayne. 2009. "Luck, Knowledge, and Control." In *Epistemic Value*, edited by Adrian Haddock, Alan Millar, and Duncan Pritchard, 331–38. Oxford: Oxford University Press.

Sosa, Ernest. 2007. *A Virtue Epistemology: Apt Belief and Reflective Knowledge*, volume 1. Oxford: Oxford University Press.

———. 2011. *Knowing Full Well*. Princeton: Princeton University Press.
Statman, Daniel. 1991. "Moral and Epistemic Luck." *Ratio* 4:146–56.
Turri, John. 2011. "Manifest Failure: The Gettier Problem Solved."
Philosophers' Imprint 11, no. 8:1–11.
———. Forthcoming. "Unreliable Knowledge." *Philosophy and Phenomeno-
logical Research*.

CHAPTER 7

SUBJECT-INVOLVING LUCK

JOE MILBURN

1. Introduction

In recent years, philosophers have tended to think of luck as being a relation between an event (taken in the broadest sense of the term) and a subject.[1] According to these philosophers, to give an account of luck is to fill in the right-hand side of the following biconditional: an event *e* is lucky for a subject S if and only if ___. We can call such accounts of luck subject-relative accounts of luck, since they attempt to spell out what it is for an event to be lucky relative to a subject.

In this essay, I want to argue that we should understand subject-relative luck as a secondary phenomenon. What is of philosophical interest is giving an account of *subject-involving luck*, i.e., filling in the right-hand side of this biconditional: it is a matter of luck that S φs iff ___.

I take it that this view is contrary to nearly all philosophers who are currently interested in the topic of luck. Most of these philosophers are, to use Nathan Ballantyne's word, *interventionists* (Ballantyne 2014). They believe that by giving an account of luck in general, we will be able to intervene in a number of philosophical issues. Ballantyne, on the other hand, has recently pushed an anti-interventionist line (Ballantyne 2011, 2014). In particular, he has forcefully argued that understanding luck will not help us understand knowledge or Gettier problems. As a result, the topic of luck fails to have the broad philosophical significance many philosophers believe it does.

---

[1] See, for example, Statman (1991), Rescher (1995), Latus (2003), Pritchard (2005), Coffman (2007), Riggs (2009), Steglich-Petersen (2010), Levy (2011), and Ballantyne (2011).

The Philosophy of Luck, First Edition. Edited by Duncan Pritchard and Lee John Whittington.
Chapters © 2014 Metaphilosophy LLC and John Wiley & Sons Ltd, except for "Luck as Risk and the Lack of Control Account of Luck" © 2015 Metaphilosophy LLC and John Wiley & Sons Ltd. Book compilation © 2015 Metaphilosophy LLC and John Wiley & Sons Ltd.
Published 2015 by John Wiley & Sons Ltd.

I agree with the majority of philosophers working on the topic of luck today (and disagree with Ballantyne) that luck is an important topic, one capable of illuminating a wide range of philosophical issues, especially those in epistemology. But I agree with Ballantyne (and disagree with the majority) that giving an account of what it is for an event *e* to be lucky for an agent S is philosophically unimportant. The problem, I argue, is that most philosophers have overlooked the difference between subject-relative and subject-involving luck. It is this latter phenomenon that is philosophically important.

If my argument works, then we should shift focus from giving an account of subject-relative luck to giving an account of subject-involving luck. One of the upshots of refocusing our energies is that *lack of control accounts of luck* (LCALs) become much more attractive. In particular, when we focus on giving an account of subject-involving luck, a wide range of classic counterexamples to lack of control accounts of luck fall by the wayside.

My essay proceeds as follows. In section 2, I explain the difference between subject-relative luck and subject-involving luck. In section 3, I argue that we should understand epistemic and moral luck as instances of subject-involving luck, not subject-relative luck. The upshot of this argument is that philosophers should be primarily interested in giving an account of subject-involving luck. In section 4, I argue that, while LCALs for subject-relative luck fall prey to counterexamples involving regularly occurring events that are beyond our control, LCALs for subject-involving luck do not face this problem. Finally, in section 5, I consider two objections to my argument and respond to them.

2. Subject-Relative Luck Versus Subject-Involving Luck

Consider the following expressions:

> It was lucky for Seth McFarlane that he missed his flight on September 11th.
> It was unlucky for Christopher Walken that he turned down the role of Han Solo.
> It was unlucky for Philipp II, but lucky for Queen Elizabeth's subjects, that a storm destroyed the Spanish Armada.[2]

Such expressions ascribe what can be called subject-relative luck. Accounts of luck in the current literature tend to be accounts of subject-relative luck. According to these accounts, luck is a property of an event in relation to an individual. These accounts of luck then seek to elucidate this property by filling in the right-hand side of the following biconditional:

> An event *e* is lucky for a subject X iff ___.

---

[2] This last example comes from Rescher (1995, 20).

It should be noted that expressions attributing subject-relative luck are not the only "luck" expressions in common use. Other common expressions include the following:

> Seth McFarlane was lucky that he missed his flight on September 11th. (ascription of luckiness)
> It is a matter of luck that Seth McFarlane missed his flight on September 11th. (ascription of subject-involving luck)

It is plausible that expressions that ascribe luckiness are not importantly different from expressions that ascribe subject-relative luck. In the one, luck is understood as a property of an event in relation to a subject; in the other, luck is understood as a property of a subject in relation to an event. The following biconditional seems to hold:

> E is lucky for S iff S is lucky that E.

In this case, if it was lucky for Seth McFarlane that he missed his flight on September 11th, then it follows that McFarlane was lucky that he missed his flight on September 11th, and vice versa. The difference between asserting one or the other of these statements is insignificant.

But this is not true of expressions that ascribe what I am calling subject-involving luck; it seems that these are importantly different from expressions that ascribe subject-relative luck. First, on a natural way of understanding things, subject-involving luck is not a property of an event in relation to a subject, or the property of a subject in relation to an event. Rather it is the property of an event that involves a subject. What is lucky is simply this: That S φs (for example, that Seth McFarland misses his flight). Because of this it could be a matter of luck that Seth McFarlane missed his flight on September 11th, while it was neither lucky nor unlucky *for* McFarlane that he missed his flight.

This point becomes clearer when we consider what Ballantyne (2011, 2012) calls the "significance condition" on subject-relative luck. Ballantyne points to unanimous agreement among philosophers working on luck that for an event *e* to be lucky for S, *e* must somehow be significant for S. Exactly what it is for an event to be significant to a subject S is open to debate.[3] However, that there is some significance condition for subject-relative luck is highly intuitive.

It is important to note that there is no significance condition for subject-involving luck. Perhaps missing his flight is a matter of complete indifference for Seth McFarlane; his subjective and objective well-being remains the same whether or not he misses the flight. Still, it could be asked why

---

[3] See Ballantyne (2012) for some of the moves in this debate.

McFarlane missed his flight, and it could be truly said in response that it was merely a matter of luck that McFarlane did so.

Or consider the following example. Suppose that the artist in residence at your university conducts an Absurdist Raffle as a work of performance art. He assigns every student and faculty member in the university a number, puts these numbers in a very large hat, and draws one. To "reward" the winner, the artist gives a member of the university administration $100,000 in Monopoly money.

Suppose that you are the winner of the Absurdist Raffle, but that you never find out about it. Surely it is neither lucky nor unlucky for you that you won the raffle. After all, we can suppose that winning the Absurdist Raffle is and should be completely insignificant to you. But while it is neither lucky nor unlucky for you that you won the raffle, it does seem to be a matter of luck that *you* won the raffle. Indeed, it is a paradigmatic example of something that happens as a matter of luck. So for subject-involving luck, unlike for subject-relative luck, there is no significance condition.

3. Epistemic Luck and Moral Luck Are Instances of Subject-Involving Luck

In the last section I made a distinction between subject-relative luck and subject-involving luck. This distinction is important to note when one considers why philosophers are currently interested in luck.

Take for instance Duncan Pritchard's anti-luck epistemology (Pritchard 2005, 2007). Pritchard's program can be understood as starting from a basic platitude, namely, "knowledge excludes luck," and building up from this platitude an account of knowledge. Doing this, says Pritchard, involves three steps: "First, we offer an account of luck. Second, we specify the sense in which knowledge is incompatible with luck. Finally, third, we put all this together to offer an anti-luck analysis of knowledge" (2007, 278). So Pritchard's account of luck is given with the ultimate aim of providing an account of knowledge.

Or consider Neil Levy's (2011) sustained argument against our being morally responsible agents. Again, Levy's program is to start with a basic platitude, that luck excludes moral responsibility, and to build up from this platitude some substantial philosophical thesis, namely, that we are never morally responsible. So Levy also has ulterior motives, so to speak, in giving a general account of luck. And by and large these cases generalize. I am aware of no contemporary philosopher who gives an account of luck who is not motivated directly or indirectly by issues involving epistemic or moral luck.

One question that is worth asking, then, is this: Is it subject-involving or subject-relative luck that is relevant for debates concerning moral or epistemic luck? I think the answer is clear: it is subject-involving luck. It follows that philosophers should be primarily concerned with giving an account of subject-involving luck.

Let's start with epistemic luck. Consider a paradigmatic example of epistemic luck: Gettierized belief.

Ballantyne (2011) shows us that understanding epistemic luck as an instance of subject-relative luck creates a number of difficulties. Consider the following example:

> Green and Blue are … looking over the field, gazing at what appears to be a sheep. They form true beliefs that a sheep is in the field. Though a sheep is just out of view, what Green and Blue see is a dog masquerading as a sheep. Suppose that the very same epistemic facts hold for Green and Blue, but that different significance facts hold for them. Having a true belief regarding that sheep is good for Green. … Not so for Blue. He is apathetic and … deeply depressed. In fact, suppose that Blue has no preference or desire to have a true belief that a sheep is in the field; he just doesn't care. So, having a true belief about the sheep is neither good nor bad for him. (Ballantyne 2011, 493–94)

If we allow that it is insignificant for Blue that he believes truly that there is a sheep in the field, then if we understand epistemic luck as an instance of subject-relative luck, it follows that Blue's Gettierized belief is not subject to epistemic luck. Now according to anti-luck epistemologists, one fails to know when one's belief is Gettierized, because one's belief is subject to epistemic luck. So, given anti-luck epistemology, if epistemic luck is subject-relative, then it seems that either Blue must be said to have knowledge or anti-luck epistemology fails to explain why subjects in Gettier cases fail to have knowledge. In either case anti-luck epistemology is a failure.

One way of responding to Ballantyne's counterexample is to understand the significance condition on luck in terms of ideal conditions, such as conditions of full rationality. In this case, an event e will be significant to an agent S just in case, in conditions of full rationality, S would care whether or not e occurred. One might then concede that, while Blue actually doesn't care whether or not he believes the truth that there is a sheep in the field, if he were in conditions of full rationality he would care to believe the truth of the matter. So Blue still meets the significance condition on luck, and it is correct to say that it is lucky for Blue that he believes truly that there is a sheep in the field.

This move fails to fully solve the problem for subject-relative accounts of epistemic luck. It seems that, even in conditions of full rationality, one can be indifferent to whether one believes extremely trivial truths.[4] For instance, a stranger on the bus tells you that it's his great-great-great grandfather's birthday. In conditions of full rationality, must one care whether one believes this truth? If the correct answer is yes, this isn't obvious.

More importantly, we can still construct scenarios in which it should be insignificant to us that we believe some truth, even if we accept that in conditions of full rationality we will care, all things being equal, to believe

---

[4] See Ballantyne (2011) for a related discussion.

the truth that p, regardless of what the content of p is. Consider the following case. Suppose that if we believe the truth that there is a sheep in the field before us, we will be kept from believing an equally important truth by some evil demon. It follows that our overall epistemic standing will be the same regardless of whether we believe that there is a sheep in the field in front of us. In this case, it seems that even in conditions of full rationality we should be indifferent to whether or not we believe truly that there is a sheep in the field in front of us.

This allows us to construct a variation on Ballantyne's Blue case that spells trouble for understanding epistemic luck in terms of subject-relative luck. Consider the following case.

Revised Blue: All of the facts of the story hold as above, except for the following. First, it is an objective good for Blue to believe the truth, regardless of what this truth may be. Second, it is not only true that there is a sheep in the field, but it is also true that there is a goat in the field. And third, there is an evil demon who will see to it that Blue will fail to believe the truth that there is a goat in the field if he believes truly that there is a sheep in the field, and vice versa.

In this case, all things considered, it is neither good nor bad for Blue to believe truly that there is a sheep in the field. It follows that it is neither lucky nor unlucky for Blue that he believes truly that there is a sheep in the field. Still, Blue's belief that there is a sheep in the field is subject to epistemic luck. His belief is Gettierized.

Pace Ballantyne, the upshot of these reflections should not be a rejection of anti-luck epistemology. Rather, it should be that we do not try to understand epistemic luck as a species of subject-relative luck. We should understand epistemic luck as an instance of subject-involving luck. Note that, even in Revised Blue, it seems right to say that *it is a matter of luck that Blue believes truly that there is a sheep in the field.* This is because there is no significance condition on subject-involving luck. Consider again the Absurdist Raffle case. While winning the absurdist raffle lacks (and should lack) any significance for you, it is still a matter of luck that you win the absurdist raffle. Likewise, even if we suppose that in the Revised Blue case, whether or not Blue believes truly that there is a sheep in front of him is (and should be) insignificant to him, it still is a matter of luck that he believes the truth.

It follows that we should understand epistemic luck as a species of subject-involving luck. What about moral luck? Consider the following case of so-called resultant moral luck.

Drunk Driver: Sam decides to have a liquid lunch, and has way too much to drink. Seeing double, somehow he manages to start his car; he drives home on his usual route, past the local elementary school. He does not hit anyone.

This example serves as an example of resultant moral luck, because, while it is clearly beyond Sam's control whether he hits a child on his way home, his avoiding such an accident determines his moral standing. Had he hit a child, we would judge him not only for being a glutton and reckless but also for being a killer.[5]

I think considerations similar to those given above show that we should understand moral luck as being a species of subject-involving luck. First, it seems that we have said enough to imply that Sam is not in control of avoiding an accident when we say that it is a matter of luck that he did not hit anyone; second, we can construct scenarios where it is intuitively not lucky for Sam that he fails to hit anyone as he drives home but in which he is still subject to moral luck.

On the one hand, consider examples in which one is indifferent to Sam's moral standing.

Blue Drunk Driver: Sam has a liquid lunch. He is quite drunk. He makes it to his car, somehow, and drives the normal way from the bar to his home, past the elementary school as the children are being dismissed. Somehow he makes it home without hitting any children. But he is so depressed he is indifferent to having hit a child or not. To him, the world is equally bad, whether or not he is branded a killer.

In this case it, seems that Sam is neither lucky nor unlucky to have avoided hitting any children with his car. It simply doesn't matter to him.

To avoid such cases, again, one might appeal to ideal conditions, for example, conditions of full rationality. One might claim that in conditions of full rationality one must care about one's moral standing because this is somehow tied into being a fully rational agent or perhaps because having a good moral standing is an objective good for one's self. Then one might concede that, while in Sam's actual state it insignificant to him whether he hit a child, in conditions of full rationality, not hitting a child would be significant to him, and so the significance condition for subject-relative luck is met in Blue Drunk Driver.

But, again, appeals to ideal conditions will not solve the problem. Suppose that we reject *extreme moral rationalism*. Someone is an extreme moral rationalist just in case she believes that nothing can be more significant to someone (in conditions of full rationality) than the quality of her moral standing. Extreme moral rationalism is highly controversial. Once we reject extreme moral rationalism, cases of moral luck that are neither lucky nor unlucky for a subject abound. To construct such cases, make it a matter of luck that an agent S φs where φing bears on his moral standing. Then stipulate that S's φing ensures that S will ψ, and that S's ψing offsets whatever good is gained or evil suffered by S through his φing.

---

[5] See Nagel (1976) for a characterization of moral luck along these lines.

For instance, suppose that there is an absolute prohibition on lying, and that it is an objective good for everyone to tell the truth. In this case, Sam (in conditions of full rationality) prefers to tell the truth, all things being equal. But suppose that it is also just as much an objective good for Sam to not hurt his friend's feelings. Then it is just as important to Sam, in conditions of full rationality, that Sam not hurt his friend's feelings as it is for Sam to tell the truth. Suppose that Sam has been drinking to the point where he is not in control over what comes out of his mouth; his friend asks Sam if Sam thinks the friend is a talented poet; Sam speaks truly: "Not really, though I think you are a great person," thereby hurting the friend's feelings. In this case, Sam is not in control his telling the truth, though *ex hypothesi* this bears on his moral standing. So Sam is subject to moral luck. It seems, however, that Sam is neither lucky nor unlucky to have told the truth. After all, it is just as important to him not to hurt his friend's feelings as to tell the truth. In this case, moral luck is best seen as an instance of subject-relative luck rather than subject-involving luck.

Furthermore, even if one accepts extreme moral rationalism, it is still possible to construct these sorts of scenarios. All we need are events of equally moral significance to cancel each other out, and we can construct scenarios where by ϕing one is subject to moral luck, though one is neither lucky nor unlucky. Consider the following revision of Drunk Driver:

Drunk Driver*: Sam decides to have a liquid lunch, and has way too much to drink. Seeing double, somehow he manages to start his car; he drives home on his usual route, past the local elementary school. He does not hit anyone. However, this ensures that he will do some other action ψ, which is just as bad as hitting a child with a car while drunk. (For instance, not hitting a child inevitably results in Sam taking target practice with his bow and arrow in a crowded area and shooting a child.)

Here it is neither lucky nor unlucky for Sam that he avoids hitting a child with his car on the way home from the bar. In this particular scenario, not hitting a child with his car simply ensures that he will perform some other irresponsible activity that results in the injury of a child. So while it is a matter of luck that he didn't hit a child with his car, and this does affect his moral standing, it seems wrong to say that it is lucky or unlucky for him that he didn't hit a child. For while his moral standing is different from it would have been if he had hit a child, it isn't better or worse.

It follows that we should not understand moral luck as being an instance of subject-relative luck. Rather, we should understand moral luck as an instance of subject-involving luck. In the cases we canvassed above, it is a matter of luck that Sam does some action such that he is not in control of his performing the relevant action, even though this action bears on his moral standing. But the additional question of whether or not he is lucky to perform the relevant action seems beside the point. Moral luck is not an instance of subject-relative luck.

The upshot of this argument is that, in general, our philosophical efforts should focus on giving an account of subject-involving luck. The topic of luck is of interest primarily for the light it promises to shine on other philosophical topics. But we have strong reasons for believing that focusing on subject-relative luck will fail to deliver on this promise, both in the case of epistemology and in the case of ethics. Thus, we should not worry about subject-relative luck. But we have no reason for rejecting the original intuition that understanding luck can provide insight into epistemological and ethical topics. Instead of focusing on giving an account of subject-relative luck, we should focus on giving an account of subject-involving luck.

4. Subject-Involving Luck and Lack of Control Accounts of Luck

In the last section, I argued that philosophers should shift their focus from subject-relative luck to subject-involving luck. In this section I want to argue that once we focus on giving an account of subject-involving luck, then lack of control accounts of luck, or LCALs, become much more plausible.

According to LCALs of subject-relative luck, an event e is lucky for S just in case (i) e is (sufficiently) outside S's control, and (ii) e meets the relevant significance condition. It is easy to transpose LCALs for subject-relative luck so that they give an account of subject-involving luck. According to LCALs of subject-involving luck, it is a matter of luck that S ϕs just in case S's ϕing is not sufficiently under S's control. Different LCALs will give different accounts of what it is for an agent's ϕing to be under her control. I believe that the most promising line is that taken by Wayne Riggs (2009). Riggs suggests that we understand control in terms of the exercise of an agent's causal powers: "One has control over some happening to the extent that the happening is properly considered something the agent has *done*. ... This imposes two separable requirements. First, the event has to be the product of the agent's powers, abilities, or skills. Second, the event has to be, at least in some attenuated sense, something the agent *meant to do*" (Riggs 2009, 214).

LCALs have been attacked in both directions. On the one hand, Jennifer Lackey has presented the following case to show that LCALs do not provide a necessary condition for subject-involving luck:

Ramona is a demolition worker, about to press a button that will blow up an old abandoned warehouse, thereby completing a project that she and her co-workers have been working on for several weeks. Unbeknownst to her, however, a mouse had chewed through the relevant wires in the construction office an hour earlier, severing the connection between the button and the explosives. But as Ramona is about to press the button, her co-worker hangs his jacket on a nail in the precise location of the severed wires, which radically deviates from his usual routine of hanging his clothes in the office closet. As it happens, the hanger on which the jacket is hanging is made of metal, and it enables the electrical current to pass through the damaged wires just as Ramona presses the button and demolishes the warehouse. (Lackey 2008, 258)

According to Lackey, it is lucky for Ramona that she blows up the abandoned warehouse: however, blowing up the warehouse seems to be more or less within her control. So LCALs fail to give necessary conditions for subject-relative luck.

Obviously, if Lackey is right, then this example would also show that LCALs fail to give necessary conditions for subject-involving luck as well. Lackey's example is convincing, however, only if one vacillates between different readings of *Ramona blows up the building*. If one distinguishes between Ramona's blowing up the building and Ramona's blowing up the building at time *t* (the exact time she blew up the building), then it seems right to say that it was a matter of luck that Ramona blew up the building at time *t*. It seems, however, that Ramona's so doing was *not* under her control. On the other hand, that Ramona blew up the building, *tout court*, seems to be under the control of Ramona, since she is a demolition expert. But understood in this way, it is not a matter of luck that Ramona blows up the building. Even if she hadn't blown up the building at time *t*, she would presumably figure out what was wrong and get the job done.

The problem for LCALs for subject-relative luck is not that they fail to provide *necessary conditions* for an event's being lucky for an agent S. Rather, the real problem is that they fail to provide *sufficient conditions* for an event's being lucky for an agent S. There are many events that intuitively are not lucky for us but are both significant for us and beyond our control. For example, intuitively it is not lucky for us that the sun will rise tomorrow. This event is obviously beyond our control, however, and presumably quite significant for us.[6] Or consider the following example from Lackey (2008, 258). It is quite significant for a child that she get fed her dinner; furthermore, it is largely beyond this child's control that she get fed her dinner. Nevertheless, given that her parents reliably feed her dinner, it is wrong to say that it is a matter of luck that she gets fed her dinner tonight. The upshot of these examples is that while simple LCALs might provide necessary conditions for an event's being lucky for an agent, they do not provide sufficient conditions.

Riggs has tried to supplement the simple LCAL account of subject-relative luck so that it might provide necessary and sufficient conditions for an event's being lucky for an agent. His account goes as follows: "E is lucky for S iff (a) E is (too far) out of S's control, and (b) S did not successfully exploit E for some purpose, and (c) E is significant to S (or would be

---

[6] For this example see Latus (2003, 467) and Pritchard (2005, 127).

significant, were S to be availed of the relevant facts)" (Riggs 2009, 220). Riggs does not explicitly define what it is for an agent S to exploit an event e, though he does provide an illustration of it. Consider, for instance, Riggs's case of Indiana Jones and New Jersey Smith.

> New Jersey Smith plans an expedition into the wilds of Africa where certain tribes of Africans with exotic customs were known to live. ... He proposes the trip to his fellow adventurer Jones, including the specific times that he means to travel. Jones agrees to tag along. As it happens, the particular tribe that lives in the area that Smith and Jones visit has a custom of sacrificing people from outside the tribe on the equinoxes of the year. The autumnal equinox happens to fall during the time that Smith and Jones are in the area, so they are captured and held until that day so they can be sacrificed. When the day of the autumnal equinox dawns, the tribe readies the captives for sacrifice at midday. As the tribesmen approach to kill them ... there is a total eclipse of the sun. The members of this tribe always take such exotic natural occurrences to signal the anger of their gods at them for whatever they happen to be doing at the moment. Consequently, they set the captives free. (Riggs 2009, 216)

Riggs has us suppose that Indiana Jones knows all along that the tribesmen would likely capture Jones and Smith, and that in response to the eclipse the tribesmen would set their captives free. Jones's knowledge of this figured into his practical plans all along. Smith, unlike Jones, is ignorant of these facts, and so cannot incorporate them into his practical reasoning.

According to Riggs, the occurrence of the eclipse is lucky for Smith but not lucky for Jones. This is because Jones, unlike Smith, exploits the occurrence of the eclipse.

> What seems to distinguish Jones from Smith, and makes Smith lucky to be alive but not Jones, is not that Jones *knew* about the eclipse and whatnot, but that he *exploited* those facts to his own advantage. That is to say, he took them into account and planned a course of action that assumed that those things would occur. And the outcome that resulted, his survival, was a consequence of his having taken account of and exploited those facts. ... The eclipse was a matter of luck for Smith because it was both out of his control and unexploited by him. The eclipse was not a matter of luck for Jones because, though it was out of his control, he nonetheless exploited its occurrence to procure his survival. (Riggs 2009, 218)

The relation of exploitation, then, seems to be something like the following:

Exploitation: S exploits an event e, just in case, (i) before e occurs, S uses the belief that e will occur in some successful bit of practical reasoning, and (ii) S is justified in so doing.

Riggs's LCAL of subject-relative luck does provide the right results in a number of cases. On Riggs's account, it is not a matter of luck for us that the sun rose this morning, since we all presumably exploited this fact in making our plans for the day. Yesterday, as we engaged in our practical reasoning we took it for granted that the sun would rise today; furthermore, we were justified in so doing. So according to Riggs's account, it is not lucky for us that the sun rose today.

While Riggs's account is an improvement upon previous LCALs of subject-relative luck, it still gets a number of cases wrong. Suppose I've been in a coma for the past ten years. Presumably, I did not exploit the fact that the sun would rise today. Still, it would not be a matter of luck for me that the sun rose this morning. Or again, take Lackey's case of a young child who is fed dinner by her parents. We might suppose that the child's practical reasoning does not extend more than several minutes into the future, and is always directed to problems that are at hand. We can suppose, then, that the child never assumes that her parents will feed her dinner in any of her practical reasoning. The child, then, fails to exploit the fact that her parents will feed her dinner. Still, it is not lucky for her that she is fed her dinner. So Riggs's account fails to provide sufficient conditions for an event's being lucky for us.[7]

Thankfully, once one turns from giving an account of subject-relative luck to giving an account of subject-involving luck, these sorts of counterexamples are beside the point. When we give an account of subject-involving luck, we are not interested in accounting for why some event is lucky for some agent. Rather, we are interested in accounting for why it is a matter of luck that some agent φs. As a result, once we focus on subject-involving luck, issues such as whether or not it is lucky for an individual that the sun rose today fall out of the picture. As a result, LCALs of subject-involving luck are more attractive than LCALs of subject-relative luck.

5. Objections

Before concluding, I would like to consider two objections.

Objection 1

You have argued that we should not understand instances of epistemic luck and moral luck primarily as instances of subject-relative luck, because there are cases of epistemic luck and moral luck that fail to meet the significance

---

[7] Given that the counterexamples to LCALs depend on events that are (1) out of our control and (2) counterfactually robust, such that they occur in all or most nearby possible worlds, couldn't LCALs be saved by adding counterfactual conditions like those that Pritchard (2005, 2007) has proposed? Adding such conditions likely will save LCALs from the claim that they fail to provide sufficient conditions for an events being lucky relative to an agent S. However, they open LCALs to the objection that they fail to provide necessary conditions for an event's being lucky relative to an agent S. So Lackey (2008) and Hiller and Neta (2007) have provided examples of events which are lucky for an agent but which are counterfactually robust. See Riggs (2009, 207–14) for further discussion.

condition. As a result you argue that philosophers should focus their energy on giving an account of subject-involving luck. But this inference might be resisted. It seems that there are instances of epistemic luck and moral luck that are not instances of subject-involving luck. Consider the following case, which is modified from Turri (2011):

> Holmes wants more than anything in the world to boost Watson's confidence as a detective. To do this, he wants to make sure that Watson forms as many true beliefs as possible, without directly intervening in Watson's thought process. Watson is currently investigating a crime scene. Given the evidence that Watson is slowly gathering, he will be justified in believing that the perpetrator of the crime has a limp. Holmes sees that Watson missed an important piece of evidence that shows conclusively that Mr. Plumb, who walks just fine, is the perpetrator. Holmes does not want Watson to get discouraged, so he runs out, finds Mr. Plumb, and kicks him in the knee, ensuring that Mr. Plumb will have a limp. Meanwhile, Watson finishes gathering the evidence and forms the true belief that the perpetrator has a limp.

Intuitively this is a case of epistemic luck. But it seems that it isn't a case of subject-involving luck. After all, given Holmes's desires and abilities, it isn't a matter of luck that Watson believes truly that the perpetrator has a limp.

Or again, consider a case in which Sam drives home hopelessly intoxicated through busy streets. Unbeknownst to him, his car can be operated by remote control by a friend of his who can observe exactly what the car is doing; the friend is quite good at controlling the car remotely; the friend will take control if Sam shows signs of being about to hit someone. Here, it is not a matter of luck that Sam avoids hitting someone. After all, his friend will see to it that this occurs. Still, Sam's not hitting anyone is a matter of moral luck.

The upshot of these examples is that you have not provided enough reason for philosophers to shift their attention from providing accounts of subject-relative luck to providing accounts of subject-involving luck. Rather, philosophers concerned in epistemic luck and moral luck should think of these phenomena as sui generis.

Reply to Objection 1

I want to argue that these instances show that there is a certain ambiguity in the phrase "it is a matter of luck that S φs." In one sense, I am willing to grant that it is not a matter of luck that Watson believes truly that the perpetrator has a limp, or that Sam avoids hitting anyone on his way home. After all, there is someone ensuring that this is the case. But in another sense, I want to say it obviously *is a matter of luck* that Watson believes the truth and that Sam arrives home safely. To accommodate these conflicting intuitions, we simply need to recognize that there is a kind of relativity in ascriptions of subject-involving luck. Relative to Watson's and Sam's powers, abilities, and skills, it is a matter of luck that Watson believes truly and Sam arrives home safely. But relative to the powers, abilities, and skills of Holmes and Sam's friend, it is not.

Objection 2

You argue that LCALs become much more attractive once we focus on subject-involving luck. Your basic argument is that counterexamples that plague LCALs of subject-relative luck don't bear on LCALs of subject-involving luck. But this doesn't seem to be true. Consider again Lackey's case of the young girl who is regularly fed dinner. Consider the following event. The young girl is fed dinner tonight. Now, given that the girl's parents regularly feed her, this event is not a matter of luck. However, it is outside the girl's control that she is fed tonight. So according to LCALs of subject-involving luck it is a matter of luck that the girl is fed tonight. Nothing has been improved.

Reply to Objection 2

LCALs of subject-involving luck attempt to explain what it is for it to be a matter of luck that S φs. "S's φing" here is properly understood as some act of S's, that is, something that S *does*. But in the case mentioned above, "getting fed" is not something the girl is properly said to do. Rather it is something that happens to her. LCALs of subject-involving luck, then, are interested in explaining what it is for it to be a matter of luck that the girl is fed, only in so much as her being fed is an act of her parents or someone else feeding her. And given that it is under her parents' control that the girl is fed, LCALs will state that it is not a matter of luck that her parents feed her.

6. Conclusion

In this essay I have argued that both epistemic luck and moral luck are best understood as instances of subject-involving luck. As a result, philosophers who believe that understanding luck in general can illuminate issues in epistemology and ethics should focus on understanding subject-involving luck. I have also argued that LCALs become more attractive once we shift our focus. Of course, more work needs to be done. On the one hand, more needs to be said concerning the relativity of ascriptions of subject-involving luck. On the other hand, more needs to be said concerning what it is for S to have control over his φing. I have only briefly touched upon these matters. I am happy, however, if I have given philosophers interested in luck reason to investigate these issues.

Acknowledgments

Special thanks to participants at the June 2013 Workshop on Luck at the University of Edinburgh, especially Lee John Whittington and Duncan Pritchard, as well as Kieran Setiya for comments on an earlier draft of this essay.

References

Ballantyne, N. 2011. "Anti-luck Epistemology, Pragmatic Encroachment, and True Belief." *Canadian Journal of Philosophy* 41:485–504.
———. 2012. "Luck and Interest." *Synthese* 185:319–34.
———. 2014. "Does Luck Have a Place in Epistemology?" *Synthese* 191: 1391–1407.
Coffman, E. J. 2007. "Thinking About Luck." *Synthese* 158:385–98.
Hiller, A., and R. Neta. 2007. "Safety and Epistemic Luck." *Synthese* 158:303–13.
Lackey, J. 2008. "What Luck Is Not." *Australasian Journal of Philosophy* 86, no. 2:255–67.
Latus, A. 2003. "Constitutive Luck." *Metaphilosophy* 34:460–75.
Levy, N. 2011. *Hard Luck*. Oxford: Oxford University Press.
Nagel, T. 1976. "Moral Luck." *Proceedings of the Aristotelian Society* 76:136–50.
Pritchard, D. 2005. *Epistemic Luck*. Oxford: Oxford University Press.
———. 2007. "Anti-luck Epistemology." *Synthese* 158:277–97.
Rescher, N. 1995. *Luck: The Brilliant Randomness of Everyday Life*. New York: Farrar, Straus and Giroux.
Riggs, W. 2009. "Luck, Knowledge, and Control." In *Epistemic Value*, edited by A. Haddock, A. Millar, and D. Pritchard, 204–21. Oxford: Oxford University Press.
Statman, D. 1991. "Moral and Epistemic Luck." *Ratio*, new series, 4: 146–56.
Steglich-Petersen, A. 2010. "Luck as an Epistemic Notion." *Synthese* 176:361–77.
Turri, J. 2011. "Manifest Failure: The Gettier Problem Solved." *Philosophers' Imprint* 11:1–11.

CHAPTER 8

THE MODAL ACCOUNT OF LUCK

DUNCAN PRITCHARD

1. Anti-Luck Epistemology and the Modal Account of Luck in Outline

The aim of this essay is to revisit the modal account of luck that I set out in previous work—especially Pritchard 2005—and defend it against objections that have recently arisen in the literature. As we will see, key to my defence of this proposal is the claim that objections to this account often ignore key features of the view. With that in mind, I want to take some time to restate the position.

The backdrop to my interest in offering a theory of luck is that I wanted to develop a way of approaching the theory of knowledge that I call *anti-luck epistemology*. It is a widely held platitude in epistemology that knowledge is in some fundamental sense incompatible with luck. Call this the *anti-luck platitude*. Until quite recently this platitude was taken largely at face value, as something that did not require further elucidation. The guiding thought behind anti-luck epistemology is precisely that one should *not* take the anti-luck platitude at face value but, rather, carefully unpack it.

In particular, anti-luck epistemology urges a three-stage approach to the theory of knowledge that takes the anti-luck platitude as central to the project. First, one offers a theory of luck. Second, one delineates the specific sense in which knowledge is incompatible with luck. Finally, third, one puts these two component parts together and formulates an *anti-luck condition* on knowledge that captures the specific sense in which knowledge is incompatible with luck. If the anti-luck platitude does reveal something deep and important about knowledge, then by undertaking the anti-luck epistemological project one should determine a core epistemic condition on knowledge. Indeed, one might even determine an epistemic condition

The Philosophy of Luck, First Edition. Edited by Duncan Pritchard and Lee John Whittington.
Chapters © 2014 Metaphilosophy LLC and John Wiley & Sons Ltd, except for "Luck as Risk and the Lack of Control Account of Luck" © 2015 Metaphilosophy LLC and John Wiley & Sons Ltd. Book compilation © 2015 Metaphilosophy LLC and John Wiley & Sons Ltd.
Published 2015 by John Wiley & Sons Ltd.

that is, with true belief, sufficient, or close to being sufficient anyway, for knowledge. That would be quite a result.

One of the attractions of this theoretical project is that it might offer a principled way of dealing with the Gettier problem—*viz.*, the problem of explaining what it takes to avoid the specific kind of epistemic luck that undermines knowledge in Gettier-style cases. On standard epistemological proposals one tries to deal with the Gettier problem by working out which condition or conditions one needs to add to one's favoured non-Gettier-proof account of knowledge in order to make it Gettier-proof. This way of approaching the problem has been notoriously unsuccessful, and tends to lead to analyses of knowledge that strike one as ad hoc and unmotivated. Indeed, the general lack of success of this way of dealing with the Gettier problem has given rise to the widespread view that knowledge is not the kind of thing that is susceptible to an analysis.[1]

Anti-luck epistemology promises to be a better way of approaching this issue since rather than focussing specifically on the kind of epistemic luck in play in Gettier cases, one instead attends to the general question of the nature of knowledge-undermining epistemic luck, whether it is found in Gettier-style cases or elsewhere. The anti-luck condition that results from a successfully conducted anti-luck epistemology will thus not be a mere anti-Gettier condition, even though it will exclude Gettier-style cases just as it excludes other cases of knowledge-undermining epistemic luck. Moreover, rather than taking the notion of luck as a primitive, anti-luck epistemology incorporates a theory of luck as a means of outlining the anti-luck condition on knowledge.

The difficulty that faced anti-luck epistemology when I first tried to develop it, however, was that I found to my surprise that there was next to nothing in the philosophical literature on the nature of luck. Indeed, I think it is fair to say that luck was at this time largely treated as an unde-fined primitive.[2] This is surprising, particularly given the tendency of ana-lytical philosophers to offer theories of just about any term of philosophical interest. After all, luck is a core notion not just in epistemology but also in a number of other philosophical domains, such as ethics (moral responsi-bility), political philosophy (just deserts), and metaphysics (causation). In

---

[1] Most famously, of course, this is the view defended by Williamson (2000). Note that we are here glossing over the issue of whether an adequate analysis of knowledge must thereby be a *reductive* analysis. My own view is that this is unnecessary, and that what we seek is rather an analysis that is *informative*. (Reductive analyses are sometimes uninformative, after all, as when they are ad hoc, and non-reductive analyses can nonetheless be informative.) For more on the methodology of epistemology, see Pritchard 2012c and 2014b.

[2] The chief exception at that time was Rescher 1995, although it should be noted that this work is not a philosophical work in the way that we would ordinarily understand that description. Note too that in claiming that luck was largely treated as an undefined primi-tive, I'm not maintaining that *nothing* was said about this notion. As we will see below, some commentators—particularly those engaged in the debate surrounding moral luck—offered what might be plausibly classed as necessary conditions for luck.

any case, since there was no existing literature on the philosophy of luck to engage with, I was faced with the task of trying to offer my own account, largely ex nihilio.

But why offer a modal account of luck? Here I was guided—and am still guided—by an insight from epistemology that I think is crucial for our understanding of luck. Consider the famous lottery example. Imagine a subject who holds a lottery ticket for a fair lottery with astronomically long odds, where the draw has been made. The ticket is a loser, but the subject has not yet heard the result and so has no inkling of this. We can add to this story in two interesting ways. In the first scenario, the subject becomes aware of the astronomical odds involved and hence on this basis forms the true belief that her ticket is a loser. In the second scenario, the subject hasn't paid any attention at all to the odds involved in this lottery. Instead, she reads the result in a reliable newspaper and so on this basis forms the true belief that her ticket is a loser.

Here is the puzzle. It seems that the subject in the first scenario doesn't know that her ticket is a loser, and yet the subject in the second scenario does.[3] Moreover, the natural explanation of why this is so is that in the first scenario the subject's true belief is a matter of luck, while in the second scenario it is not a matter of luck. The reason why this is puzzling is that if we consider the subject's bases for belief then, from a probabilistic point of view anyway, the odds in the second case are nothing like as massively in support of the truth of the subject's belief as they are in the first case. How, then, can it be that knowledge is present in the second case and not the first? Is knowledge not a straightforward function of the strength of one's evidence, probabilistically conceived?

The lottery case reminds us that an event can be modally close even when probabilistically unlikely. That is, the possible world in which one wins a lottery, while probabilistically far-fetched, is in fact modally close. The possible world in which one is leaping about with joy in one's room because one is a lottery winner is very alike to the possible world in which one is tearing one's ticket up in disgust—all that needs to change is that a few coloured balls fall in a slightly different configuration.[4]

In contrast, the possible world in which a reliable newspaper misprints the lottery result, while not so probabilistically far-fetched, is not modally close. Newspapers have a morbid fear of printing erroneous lottery results—just think of the problems that this could cause—and so have elaborate systems in place in order to ensure the accuracy of what they

---

[3] See Turri and Friedman 2014 for empirical evidence that our folk judgements about lottery cases line up with the relevant philosophical judgements. For more on the lottery problem, see Pritchard 2007b. For a very different treatment of this problem, see Hawthorne 2004.

[4] I am here characterizing possible worlds in the standard way—as set out in seminal work by Lewis (1973; 1987), amongst others—in terms of a similarity ordering. I comment on some of the philosophical issues raised by possible worlds below.

print in this regard. It follows that one needs to change quite a lot about the actual world in order to get to the possible world where a reliable newspaper (the *Times*, say) prints an incorrect lottery result. The moral is that modal closeness comes apart from probabilistic closeness.[5] In particular, one cannot infer from the fact that an event is probabilistically unlikely (such as a lottery win) that it is therefore also modally far-off.

Indeed, this is why people play lotteries, and yet do not place bets on modally far-fetched events with similarly massive odds. The odds of my winning the hundred-metre gold medal at the next Olympics may well be in the region of your average lottery win, but you'd be crazy to bet the same amount you'd spend on a lottery ticket on this event obtaining. This is because not only is this event probabilistically far-fetched, it is also modally far-fetched—an awful lot would need to change about the actual world to make it such that I am an Olympic sprint champion (indeed, I suspect it would take some sort of global conspiracy).[6]

This distinction between modal and probabilistic closeness of events may seem highly theoretical, but it is nonetheless rooted in our ordinary judgements. The lottery example is a case in point, since we are surely sensitive to fact that the lottery-win scenario is modally close even while being probabilistically far-fetched (even if we wouldn't articulate this distinction in these terms of course). Indeed, that we recognise this point is revealed by our behaviour, in that we are not at all inclined to bet on modally far-fetched events with odds similar to the lottery.

Why is knowledge lacking in the first scenario, where the subject's true belief is based solely on the odds, but present in the second scenario, where it is based on reading the result in a reliable newspaper? Here is a perfectly natural explanation. In the first scenario the subject's true belief is just down to luck, since she could so very easily have formed a false belief (i.e., had the balls fallen in a slightly different configuration, such that she owned the winning lottery ticket). In the latter case, in contrast, the subject's true belief doesn't seem lucky at all. Given how she formed her true belief, she couldn't have easily formed a false belief, since reliable newspapers tend to

---

[5] As it happens, I have first-hand experience of this point about newspapers. In my late teens I gained work experience with a local newspaper and saw for myself the lengths it went to in order to ensure the accuracy of its lottery results. (And note that this is just a local newspaper with limited resources, rather than an internationally respected national newspaper like the *Times*.)

[6] Note that the slogan for the United Kingdom's national lottery is: "It could be you!" This is clearly not the "could" of probability, since in this sense it (realistically) couldn't be you, but rather the "could" of modal nearness—i.e., if you play the lottery, then someone just like you will win it. This is borne out by the advertising campaign, which at one point featured a God-like finger hovering over ticketholders, and then zapping one of them (the winner). Note that in arguing that one would be crazy to bet on a modally far-fetched event with similar odds to a lottery win I am not thereby suggesting that playing the lottery is rational. The point is rather that whatever one thinks of the rationality of playing the lottery, placing a bet on a modally far-fetched event with similar odds would be, from a rational point of view, much worse.

publish the right result in worlds like the actual world. We can also think of this in terms of the notion of risk. Given how the agent forms her belief in the first case it seems subject to an undue degree of epistemic risk, since she could have easily got things wrong. Not so in the second case, where the degree of epistemic risk is far lower.

Our judgements about knowledge are thus sensitive to the modal closeness of error, as opposed to its probabilistic closeness. That is, the moral of the lottery example is that a true belief can fail to be knowledge even despite the odds being massively in its favour so long as the possibility of error is nonetheless modally close. In such cases we judge the agent's cognitive success to be too risky to count as knowledge.

This point about this distinction between modal and probabilistic closeness being rooted in our everyday judgements is borne out by the empirical research on luck and risk ascriptions. For while I found, back when I first started working on anti-luck epistemology, that philosophers hadn't said much about luck, I discovered that psychologists had said a great deal about this topic. Moreover—and this is a key point—when it comes to judgements about luck and risk, it is the modal closeness of the target event that has the whip hand.

For example, in a series of studies conducted by the psychologist Karl Teigen (1995; 1996; 1997; 1998a; 1998b; 2003) it was found that when a success was perceived as being physically close to a failure (i.e., when a wheel of fortune stopped in a winning sector but was physically close to stopping in a losing sector), the success was perceived as more lucky than when the failure was not perceived as physically close. Moreover, Teigen also found that this counterfactual closeness could not be understood simply in terms of the probabilities involved. Subjects were willing to treat events as being different as regards the degree of luck involved even while granting that the probabilities of each of the two events occurring was the same. Subjects would, for example, recognise that the probability of one's ball landing in a losing sector on a roulette wheel was constant wherever the ball landed in that losing sector, while also regarding an event in which one's ball landed near to the winning sector as involving bad luck, unlike other events where the ball landed further away (which, depending on where the ball landed, were either not regarded as unlucky at all or else regarded as involving less bad luck).[7]

With these kinds of considerations in mind, we are thus closing in not just on the idea that luck is a distinctively modal notion but also on what kind of modal notion it is. That is, roughly, what makes an event lucky is that while it obtains in the actual world, there are—keeping the initial conditions for that event fixed—close possible worlds in which this event

_____

[7] Other studies confirm Teigen's findings. See especially Kahneman and Varey 1990, Tetlock 1998, and Tetlock and Lebow 2001. I survey the psychological work on luck in Pritchard and Smith 2004.

does not obtain. So, for example, a lottery win is a lucky event because there are close possible worlds where the initial conditions for this event are the same but where one does not win the lottery (i.e., where the coloured balls fall in a slightly different configuration).

What is meant here by the "initial conditions for the event"? The point of this restriction is that we need to keep certain features of the actual world fixed in our evaluation of the close worlds. In particular cases, it is usually pretty clear what needs to remain fixed. In the lottery case, for example, we obviously need to keep fixed that the subject buys a lottery ticket and that the lottery retains many of its salient features (i.e., remains free and fair, with long odds, and so on). If, say, one were *guaranteed* to win the lottery (e.g., it is rigged in one's favour), then clearly this isn't a lucky event even if, as it happens, there are close possible worlds in which one does not win the lottery (e.g., because one is prevented from buying a lottery ticket in such worlds).[8]

Is there a general specification that one can offer of these "initial conditions"? Well, we can say this much: they need to be specific enough to pick out a particular kind of event that we want to assess for luckiness, but not so specific as to guarantee that this event obtains (e.g., we don't want the purchase of a *winning* lottery ticket to be part of the initial conditions for the lottery win). That's quite vague, of course, but my suspicion is that we shouldn't expect anything more detailed, in that we shouldn't require a theory to be any more precise than the phenomena about which we are theorizing. For our purposes it is enough that we can pick out such initial conditions on a case-by-case basis (which I believe we usually can).

Note that this conception of luck can accommodate the idea that luck comes in degrees. Consider the lucky event of not being shot by a sniper's bullet. With everything else kept fixed, imagine that in scenario A the bullet misses one by millimetres, whereas in scenario B the bullet misses one by a metre. We would naturally judge that both events are lucky, and our account of luck confirms this judgement, for in both cases there are close possible worlds in which the initial conditions for this event are the same and one is hit by the bullet. We would also naturally judge that scenario A is luckier than scenario B, and our account of luck again confirms this judgement. For the possible worlds in which one is hit by the bullet are clearly closer in scenario A than they are in scenario B, since less needs to be changed about the actual world in order to get to these possible worlds.

---

[8] Of course, it may be lucky in such a case that one gets to buy a lottery ticket, but that's a different event from the one under consideration, which is one's winning of the lottery. (This point reminds us of the importance of making the target lucky event clear and keeping it fixed throughout our evaluation.) Relatedly, the agent concerned might still regard the lottery win as lucky, but this would be because he or she is not in possession of relevant information about this event. As I explain below, we are interested in luck as an objective phenomenon, and not merely in subjects' judgements about luck, regardless of their epistemic pedigree.

More generally, we can say that the degree of luck involved varies in line with the modal closeness of the world in which the target event doesn't obtain (but where the initial conditions for that event are kept fixed). We would thus have a *continuum* picture of the luckiness of an event, from very lucky to not (or hardly) lucky at all. Once the degree of luck falls below a certain level—i.e., once there is no modally close world where the target event doesn't obtain—then we would naturally classify the event as *not* lucky, since it does not involve a significant degree of luck. This conception of luck also allows us to compare events in terms of their luckiness, so even when considering two events that we don't think are lucky we can nonetheless ask the question whether the one is luckier than the other (e.g., not being shot by a sniper is presumably luckier if the actual world is one where gun ownership is common than if, all other things being equal, gun ownership is rare—even when neither event involves a significant degree of luck).[9]

Another way of thinking about luck is in terms of the related notion of risk. There is a lot of empirical support for the idea that subjects' judgements about risk and luck tend to go hand in hand. In particular, just like luck, subjects' judgements about risk track the modal closeness of the target event rather than its probabilistic likelihood. So subjects might grant that the probabilistic likelihood of two events is broadly the same, and yet nonetheless characterize one of them as being riskier than the other because they regard this event as modally closer.

A good example of this is subjects' judgements about the risks involved in various kinds of transport. While subjects will grant that the probability of sustaining serious injury when, say, driving a car is much, much higher than when using alternative forms of transport, such as taking the train, they nonetheless tend to judge that car driving is not an especially risky activity (i.e., no more risky, or at least not especially riskier, than taking the train). There are various explanations for this. It is certainly true, for example, that various cognitive biases have a role to play in leading subjects to make these assessments of risk. The fact that one is driving one's car, as opposed to being a passenger (as on a train), makes the "illusion of control" bias relevant, for example.[10] This leads subjects to overestimate their control over events associated with car driving, such as their propensity to have accidents. This bias, coupled with the fact that subjects tend to overestimate their expertise (most people think that they are above-average

---

[9] See Kahneman and Varey 1990 and Teigen 1996 for discussion of how subjects' judgements about degrees of luck (/risk) vary in proportion to the counterfactual closeness of the target event.

[10] Indeed, subjects tend to judge travelling by car as more risky when it is made clear that they will be a passenger rather than the driver. See, for example, McKenna 1993. For more on the illusion of control, see Langer 1975 and Thompson 1999 and 2004.

drivers),[11] leads them to regard driving a car as a not especially risky activity, even when taking into account the relatively high probability of car accidents (when compared with some other forms of transport).[12]

We can explain what is going on here in terms of our account of luck. The car driver in the grip of these cognitive biases has a conception of the actual world such that the possible world in which he or she incurs serious injury while driving is not especially close, and hence not a serious risk (even despite the relatively high probabilities in play). That is, keeping fixed salient initial conditions (the subject's above-average driving skill, for example), it is not a matter of luck that he or she avoids a serious accident on a given car journey. Judgements about luck thus dovetail with judgements about risk. To say that a target event is risky is to say that (keeping relevant initial conditions for that event fixed) it obtains in close possible worlds. As the modal distance between the actual world and the possible world where the target event obtains becomes more remote, so the riskiness of the event lessens. At some point, the target event is so modally remote as to not be significantly risky, and hence we tend to judge that there is no risk involved.

Of course, there are some differences between our judgements about luck and about risk. The latter specifically concerns "negative" events, such as hazards (e.g., the risk of being in a car accident), while we can quite comfortably think of the former in terms of positive events like lottery wins. But at least in so far as we are concerned to eliminate luck from an event—which usually means eliminate *bad* luck—the two terms will tend to coincide.[13] This is certainly true in the epistemic case, where our concern is to find a condition on knowledge that excludes the negative event of bad (i.e., knowledge-undermining) epistemic luck.[14]

Let us now return to anti-luck epistemology. We now have the theory of luck that we are looking for, in the form of the modal account of luck. In terms of the specific sense in which knowledge excludes luck, let us gloss over the dialectical twists and turns, of which there are many, and cut to the chase: we are interested in the event of the subject being cognitively

---

[11] This is the so-called *overconfidence bias*. Famously, in a U.S. study Svenson (1981) found that 93 percent of drivers rated their driving abilities as above average. Indeed, interestingly (though perhaps not surprisingly), those with low levels of skill are often *more* apt to overestimate their skill levels, a phenomenon known as the "Dunning-Kruger effect." See Kruger and Dunning 1999.

[12] Does it matter that we are dealing here with luck/risk ascriptions that are the product of cognitive bias? As I explain below, I don't think it does.

[13] I discuss the notion of risk and how it relates to luck in more detail in Pritchard 2014a.

[14] That we are offering an account of luck *simpliciter*, and not of good or bad luck in particular, requires emphasis. Whether the luck is good or bad is a further judgement that we bring to bear on the event, in terms of whether it is positive or negative. I don't think it should be part of a theory of luck to say much more about good and bad luck than this, for it is luck *simpliciter* that we are interested in. I say more about this point in section 2.

successful (i.e., having a true belief), and we want this event to be non-lucky.[15] With our modal account in mind we can flesh this out by saying that a lucky cognitive success is a cognitive success where—keeping the relevant initial conditions fixed as usual—cognitive failure (i.e., false belief) is modally close.[16] A non-lucky cognitive success is thus one where, keeping the initial conditions fixed, cognitive failure is not modally close. A bit more carefully, we can say that there is a continuum of epistemic luck in play here. Where the cognitive failure in question is modally very close, then the cognitive success is very lucky and hence knowledge is excluded. Where the cognitive failure is modally far-off, then the cognitive success is not significantly lucky and hence is compatible with knowledge. In between, there is spectrum of degrees of epistemic luck (and hence we would expect our judgements about knowledge possession to be more secure as the degree of epistemic luck lessens).

As before, we can also express this idea in terms of the language of risk, in this case *epistemic risk*. Where cognitive failure is modally close, then there is a great deal of epistemic risk, enough to undermine knowledge. As the cognitive failure moves further out, modally speaking, however, we become more tolerant of it, to the point where eventually it becomes compatible with knowledge. In particular, where the cognitive failure is modally far-off, then the degree of epistemic risk is low enough to be discounted, and hence it isn't incompatible with knowledge.

Anti-luck epistemology generates a number of interesting theoretical consequences. One of the overarching morals of anti-luck epistemology is that we should endorse the so-called *safety principle*, which is a modal condition on knowledge (and in particular, we should endorse it over competing modal conditions on knowledge, such as the *sensitivity principle*).[17] Moreover, anti-luck epistemology motivates a particular rendering of this principle. The safety principle in outline demands a cognitive success that could not very easily be a failure. So put, the general fit with our anti-luck epistemology is obvious.

But once we opt for an anti-luck epistemology we also get a very specific rendering of the safety principle. Rather than the general claim that a safe

---

[15] Or, as it is put in the literature, we want this event to be immune to *veritic luck*. For more on veritic luck and the other kinds of epistemic luck (both benign and malignant) that one can delineate, see Pritchard 2005.

[16] Note that the notion of cognitive failure is potentially broader than false belief, in that a failure to believe the truth in appropriate circumstances can itself be a kind of cognitive failure. But we will be setting this kind of complication to one side here.

[17] For some key defences of the safety principle, see Luper 1984 (cf. Luper 2003), Sainsbury 1997, Sosa 1999, Williamson 2000, and Pritchard 2002. For some key defences of the senstivity principle, see Dretske 1970 and 1971, Nozick 1981, Roush 2005, Becker 2007, Black and Murphy 2007, and Black 2002 and 2008. For a comparative overview of the safety and sensitivity principles, see Pritchard 2008 and Black 2011.

cognitive success is one that could not very easily have been a failure, we get instead a continuum picture of epistemic risk involved, with modally very close epistemic risks incompatible with knowledge at the one end of the continuum, shading off along this continuum towards modally far-off epistemic risks that are compatible with knowledge. The result is a much more nuanced conception of the safety principle. As I've argued elsewhere, with the safety principle so understood we can deal with a range of problems that have been levelled against it.[18]

It's not my goal here to defend anti-luck epistemology, however. I mention it only to offer a sense of how the modal account of luck can be put to work within a particular philosophical project. What I want to do instead is further outline the modal account of luck, and in doing so deal with some of the objections that have been raised against it in the recent literature.

2. Luck, Significance, and Subjectivity

First off, we should note that the modal account of luck as just presented might be thought to be missing an important component. In particular, in earlier work I supplemented the modal condition on luck outlined above with a further condition, which I referred to as a "significance condition."[19] The thinking behind this condition is that there are all kinds of events that satisfy the modal condition that don't thereby seem to qualify as lucky— such as a "lucky" small avalanche in a distant and unoccupied region of the South Pole, one that will never be of any interest (or ought to be of any interest) to anyone. In short, it seems that it is only *significant* events that are in the market for luck.

The question then becomes how best to formulate this condition, and this poses various challenges. Does it suffice to meet the significance condition that a subject (*any* subject?) merely regards the target event as significant (whether rightly or wrongly), or should we opt for a more objective treatment of significance whereby we focus on those events that the subject *ought* to find significant? Do we allow for subject-relative luck, such that an event can be lucky for subject A and yet not for subject B? Do we allow purely pragmatic factors—such as what kinds of things are being discussed

---

[18] For example, I've argued that with safety properly understood as motivated by an anti-luck epistemology, one can evade a dilemma posed for safety-based theories of knowledge by Greco 2007 and also explain how even one's beliefs in necessary truths can be unsafe. Moreover, I claim that anti-luck epistemology has predictive power, as it enables us to adjudicate, in a principled fashion, between opposing responses to certain cases, in that it can highlight which details of the case are important and thereby explain why two superficially identical formulations of a given example can generate very different responses. For more on anti-luck epistemology, see Pritchard 2005; 2007a; 2012a; and 2012b. See also Pritchard 2007b; 2008; 2009; 2013; forthcoming *a*; and forthcoming *c*.

[19] See Pritchard 2005, chap. 6. See also Pritchard 2004 and 2006 and Pritchard and Smith 2004.

in a given conversational context—to determine whether an event is significant? And so on.

In earlier work I tried to steer a course through these issues (see especially Pritchard 2005, chap. 6), but I have now come to the conclusion that the very idea of adding a significance condition to the modal account of luck is wrongheaded. Think again about the small avalanche on the South Pole that was just mentioned. Of course, no one will regard it as lucky since no one cares about it, and it makes no difference to anyone. But why should that prevent this event from being a genuinely lucky event? The point I am getting at is that we shouldn't expect an account of the metaphysics of lucky events to be responsive to such subjective factors as whether an event is the kind of thing that people care about enough *to regard as* lucky. That's just not part of the load that a metaphysical account of luck should be expected to carry.[20]

There is a related issue in the vicinity regarding the distinction between good and bad luck. Our practices of luck ascription obviously distinguish between the two, and so one might antecedently expect a theory of luck to incorporate an account of this distinction. But as with the significance condition, I think it would be a mistake to try to build a distinction between good and bad luck into a metaphysical account of the nature of a lucky event. This distinction instead concerns our subjective responses to lucky events, and is not an inherent feature of lucky events themselves. More generally, our interest ought to be in luck as an objective feature of events, which means that we should be wary about drawing too many conclusions from agents' subjective judgements about luck.

With this point in mind, consider again the way in which we appealed to the cognitive science literature regarding luck and risk ascriptions above. One interesting feature of this literature is that it can lend support to the modal account of luck even when the judgements in question are manifestly false (e.g., where those judgements are due to cognitive bias). Consider an agent who judges that plane travel is very risky because he or she has a faulty conception of the actual world such that very little would need to occur for one's plane to crash. That the agent makes judgements about luck and risk in this way lends support to the modal account of luck even though,

---

[20] This has implications for anti-luck epistemology, because with the significance condition as part of one's theory of luck, one is in danger of endorsing pragmatic encroachment about knowledge by default (on account of the fact that the significance condition seems to bring in purely pragmatic factors). But whatever truth there may be in the pragmatic encroachment thesis, we should be wary about such a thesis simply falling out of the application of the theory of luck to one's epistemology. Hence it is an advantage to anti-luck epistemology if we can construe it in such a way that it doesn't incorporate a significance condition on luck and therefore doesn't lead to pragmatic encroachment. For further discussion of pragmatic encroachment, see Fantl and McGrath 2007 and 2011. See also Pritchard (forthcoming *b*). For further discussion of anti-luck epistemology and pragmatic encroachment, see Ballantyne 2011 and 2012.

since the agent has false beliefs about the nature of the actual world, the event in question (surviving a given plane journey) is not in fact lucky at all.

We thus find the cognitive science literature motivating a conception of lucky events that generates an extension for such events that is very different to the extension implied by our everyday judgements about luck and risk (in so far as those judgements are taken completely at face value at any rate). I don't think this should surprise us at all, any more than it should surprise us that our judgements about when an event is lucky could prove to be undermined when we are apprised of more information about the event. For example, one might regard one's lottery win as lucky so long as one takes oneself to be playing a standard lottery game, but then discover otherwise. If one subsequently discovered that one was guaranteed to win, for instance—that the game has been rigged in one's favour, say—then one would surely no longer regard it as a lucky win.

The point is that as philosophers our interest is ultimately not in our subjective judgements about luck as such (which may be made while in possession of incomplete information) but rather in luck as an objective phenomenon. More precisely, we are interested in our subjective judgements about luck only because of what they reveal about our folk concept of luck, but it is consistent with this approach that our subjective judgements about lucky events are regularly mistaken. Indeed, if the cognitive science literature in this regard is correct, then this is more than just a theoretical possibility, in that this literature reveals that we often misclassify events as lucky or not lucky.[21] As we will see below, this observation that our subjective judgements about luck are not to be taken at face value, but rather evaluated relative to an objective standard for lucky events, is important to understanding how the modal account of luck can deal with certain cases that have been levelled against it.

Could it turn out that there is no such thing as lucky events? Well, this is certainly not ruled out by the modal account of luck any more than it is ruled out by other theories of luck. If metaphysical determinism is true, for example, then one could argue that there is never any close possible world where the target event failed to obtain. But notice that it is not part of the task of offering a theory of luck that one thereby shows that there are lucky events. Indeed, it is no more a constitutive part of the task of offering a theory of luck that one demonstrates that there are lucky events than it is a constitutive part of the task of offering a theory of knowledge that one demonstrates that radical scepticism about our knowledge of the external world is false. What offering a theory of luck does entail, however, is that conditions are laid down relative to which the question of whether there are lucky events potentially has an answer.

---

[21] There has been some recognition of this point in the debate about moral luck. See Domsky 2004, Royzman and Kumar 2004, and Enoch and Guttel 2010.

3. Modality and Luck

A concern one might have about the modal account of luck relates to modality itself, especially where possible worlds are ordered in terms of similarity to the actual world in the way that is crucial to this proposal. After all, there are various problems that notoriously afflict this approach. Even setting aside concerns about the metaphysics of possible worlds, there are problems with the ordering itself. For example, it has been argued that there is no unique closest possible world to the actual world and also that there need be no fact of the matter regarding which of any two given possible worlds is closer to the actual world.[22] Should problems like these regarding possible worlds concern us?

The first point to make about this worry is to remind ourselves that the modal closeness of an event, as opposed to the probabilistic likelihood of an event, is rooted in our ordinary ways of thinking. We can glean this much from the psychological literature on luck and risk ascriptions, as noted earlier, but equally one could dip into the writings of a number of disciplines (e.g., economics, history, geography) and find examples of the very kind of counterfactual thinking and reasoning that trades on this way of thinking about modality. The point is that we *need* a similarity conception of possible worlds in order to capture what is going on in these domains, and not just in order to make sense of the notion of luck. Whatever problems such a conception faces should thus be met head-on.

The second point to make is that it's unclear whether any of the difficulties that face the similarity conception of possible worlds pose problems that are specific to the modal account of luck. Take the problem just noted that there is potentially no fact of the matter as to which of any two possible worlds is closer to the actual world, for example. Why would this problem undermine the modal theory of luck, specifically? The only way I could see it as posing a serious difficulty for this theory would be if we expected the modal account of luck to offer us a very fine-grained way of distinguishing between lucky and non-lucky events, as if there is some kind of sharp cut-off between the two. But why would we expect that (or even actively seek that in a theory of luck)? I noted earlier that luck comes in degrees, with very lucky events shading off along a continuum of luck into events that aren't significantly lucky. Such a picture is compatible with a broad penumbral range of cases where it's hard to say whether an event counts as lucky or not. Indeed, such a coarse-grained conception of luck seems entirely in keeping with our normal ways of thinking about luck. In general, there seems no reason why we would seek to appeal to possible worlds to make any fine-grained distinctions regarding lucky and non-lucky events.

---

[22] These are sometimes known as the "world border" and "world order" problems, respectively. For discussion of these problems, see Lewis 1973 and 1987.

So while it is undoubtedly the case that there are general concerns that one can raise about the particular appeal to possible worlds made by the modal account of luck, it doesn't seem to me that these concerns are in any way specific to this account, at least in so far as we understand this proposal, as I suggest we ought, as offering a coarse-grained way of individuating lucky and non-lucky events.

4. Luck and Neighbouring Notions

There is still one more feature of the modal account of luck that we need to consider before we can consider putative counterexamples to the view and rival proposals. This is that a key element of the modal account of luck is the way that it distinguishes luck from neighbouring notions like chance, fortune, lack of control, and so on. Here I will focus on the differences between luck and the neighbouring notions of chance, fortune, and accident. In the next section I will look at the relationship between luck and lack of control.

The empirical work on luck ascriptions is helpful to the modal account on more than one front here. For instance, this work shows that subjects' judgements about luck come apart from their judgements about chance. In particular, in games of chance not all "chancy" outcomes are described as lucky. That one's number doesn't come up is down to chance. But if one's number very nearly came up, then this is attributed to (bad) luck (see Wagenaar and Keren 1988 and Wagenaar 1988). This accords with the modal account of luck, in that our judgements about luck, but not about chance, are tracking what is going on in relevant regions of the modal environment. Moreover, given the point made earlier about how our judgements about luck are tracking the modal closeness of certain possibilities, rather than the probabilities associated with the target event, this is just what we should expect. "Chanciness" relates to the latter, luck to the former.

The empirical literature also marks a distinction between luck and fortune. While the former tends to be associated with particular events, the latter tends to be concerned with relatively long-standing and significant aspects of one's life, such as one's good health or financial security (see Teigen 1996 and 1997). While subjects have little tendency to characterize particular lucky events in terms of the language of fortune,[23] there is a tendency to characterize long-standing and significant features of one's life,

---

[23] The exceptions are those particular lucky events that involve large financial gains, such as lottery wins. Winning a fortune is not the same as being fortunate, however, and once we keep this distinction in mind then the temptation to think of particular lucky events—even those involving large financial gains—in terms of the language of fortune subsides. For example, we are not tempted to describe a lottery win that results not in a financial gain but in some other benefit—say, to have a prominent building named after one—in terms of the language of fortune. See also footnote 28.

which are most often described in terms of fortune, in terms of the language of luck (e.g., "I am lucky/fortunate to have such a wonderful family"). Even despite this overlap in usage, the empirical literature reveals that they are distinct notions, and hence we need to keep them apart.[24]

Indeed, notice that where one is concerned with long-standing features of one's life the language of luck is not nearly as natural as that of fortune, particularly once one makes it explicit that these features lack the characteristics at issue in the modal account of luck. Take the "luck" that one has good health, for example, and bear in mind the point made in section 2 above that we are ultimately interested in luck as an objective phenomenon (i.e., rather than our subjective judgements about luck). If one has good genes (from a health point of view) and one takes good care of oneself (one exercises, eats the right things, receives regular medical checkups, and so on), then is it really a matter of *luck* that one has good health? Surely not. We are here making it explicit that there are no close possible worlds where one's health is poor, but in doing so we are also undermining the idea that one's good health is to be characterized in terms of luck. This lends support to the modal account.[25]

A third notion that is sometimes equated with luck is that of an accident.[26] But here too we should resist the equivalence. I may take great care in choosing my numbers for the lottery and in ensuring that I purchase my ticket in time. If I win, this will be a matter of luck, but it won't be an accident. After all, I was trying to win, and took the relevant steps to make such a win possible. That precludes this event from being an accident.

Carefully distinguishing luck from neighbouring notions is crucial both to explaining how the modal account of luck is not subject to certain kinds of counterexamples and to explaining why competing accounts of luck are problematic, as we will now see.

5. The Modal Account of Luck and Its Rivals

We can finally now consider some rival accounts of luck and some putative counterexamples that have been levelled against the modal account

[24] Note that luck and fortune come apart in other ways too. For example, lucky events can be both positive (e.g., beneficial) and negative (e.g., harmful), whereas fortune is usually only used in a positive way.

[25] It is interesting to reflect on why we might use the term "luck" to describe such long-standing features of one's life. One explanation might be that it represents a kind of modesty, whereby we are disavowing any credit that might accrue to us for our good health. Another possibility is that we have a faulty picture of how our health functions such that even despite one's genes, one's diet, regular medical check-ups, and so on, there is always nonetheless a close possible world where one's health is poor.

[26] Unger 1968 offers an influential proto-anti-luck epistemology, albeit in terms of the notion of an accident rather than in terms of luck. See also Rescher 1990, though note that, as explained below, Rescher offers a different account of luck in his later work.

of luck. Consider first Nicholas Rescher's (1995) claim that lucky events are events that the subject cannot rationally expect to occur.[27] For a wide range of cases, events that satisfy the modal account of luck will also satisfy Rescher's account, and vice versa. After all, a lucky event according to the modal account of luck is an event that could very easily have not occurred. One would thus expect it to be the kind of event that one couldn't rationally expect to occur. And it is at least often the case that events that one can't rationally expect to occur, such as lottery wins, are also events that could very easily have not occurred. Crucially, however, these two accounts do come apart in terms of what they predict about particular cases, and where they do our judgements tend to go with the modal account.

For example, Rescher (1995, 35) gives the example of someone who receives a lot of money unexpectedly from a benefactor. Rescher argues that this constitutes good luck on the agent's part even if this bequest has been a long time in the planning. But I think we need to be careful here. Perhaps our agent might well be inclined to put this event down to luck, but it is not this judgement that should concern us (as it is almost certainly made while not being in possession of full information) but rather what we should say about the case given that we make all the relevant facts clear. As noted above, we are interested in whether an event is objectively lucky, which may not be the same thing as whether agents under certain conditions (e.g., where they are lacking relevant information) would judge this event as lucky.

Crucially, however, once one makes it clear that the agent knows that this bequest had been planned for a long time and was effectively guaranteed to occur, then the temptation to regard this as a lucky event subsides. Interestingly, there is a natural inclination to regard this as an example of good fortune, and I think this reflects the fact that it is (albeit unbeknownst to the agent) a long-standing and important feature of the agent's life that he will receive this money.

A related kind of example that might be thought to present problems for the modal account of luck is offered by Jennifer Lackey:

> Sophie, knowing that she had very little time left to live, wanted to bury a chest filled with all of her earthly treasures on the island she inhabited. As she walked around trying to determine the best site for proper burial, her central criteria were, first, that a suitable location must be on the northwest corner of the island—where she had spent many of her fondest moments in life—and, second, that it had to be a spot where rose bushes could flourish—since these were her favorite flowers. As it happens, there was only one particular patch of land on the northwest corner of the island where the soil was rich enough for roses to thrive. Sophie, being excellent at detecting such soil, immediately located this patch of land and buried her treasure, along with seeds for future roses to bloom, in the one and only spot that fulfilled her two criteria.

[27] A similar proposal has recently been offered by Steglich-Peterson 2010.

One month later, Vincent, a distant neighbor of Sophie's, was driving in the northwest corner of the island—which was also his most beloved place to visit—and was looking for a place to plant a rose bush in memory of his mother who had died ten years earlier—since these were her favorite flowers. Being excellent at detecting the proper soil for rose bushes to thrive, he immediately located the same patch of land that Sophie had found one month earlier. As he began digging a hole for the bush, he was astonished to discover a buried treasure in the ground. (Lackey 2008, §2)

Lackey describes Vincent's discovery of the buried treasure in this case as a "paradigmatic" instance of a lucky event, even though it is clearly not lucky by the lights of the modal account of luck given that Vincent was effectively guaranteed to be successful in finding the treasure given how the case is described. As with Rescher's example of bequest, however, I think we need to look a little more closely at the details of this case.

First off, the example is rather ambiguous in certain respects. For example, how large is this "patch of land" that Sophie locates? The example only functions as Lackey wants it to if this patch is roughly the same size as the treasure, since it is only then that Vincent is guaranteed to find the treasure in this spot. Otherwise, he could have easily planted the rose bush on this patch of land and yet not found the treasure, and that would be consistent with the event being classed as lucky according to the modal account. Moreover, how deep was the treasure buried? Presumably a treasure chest would need to be buried fairly deep to prevent it from becoming exposed accidentally (e.g., from the effects of the weather), but if that's right then it's possible that one could plant a shrub on this ground without coming across the treasure (remember, after all, that our agent is not looking for treasure). Hence there is again no obvious inconsistency between this example and the modal account of luck. And so on.

In order to remove these ambiguities, suppose we stipulated that the areas on the island where one might bury treasure all come in distinct patches not much bigger than the treasure itself, and that the soil on these patches becomes too hard to turn very quickly, so that the treasure cannot be buried very deep. Basically, we are stipulating the details of the case such that if anyone chose the patch of land in which the treasure was buried to dig (for whatever reason, including to plant a shrub), he or she would find the treasure. Now we further stipulate that there is only one patch of land on the island that is suitable for planting rose bushes, and that it is obvious that this is so to anyone who knows about these things.

With the case so redescribed, such that Vincent is guaranteed to find the treasure, are we now inclined to judge that this discovery is lucky? In particular, remember that, just as with the case that Rescher offers, we need to set aside the fact that Vincent himself might well describe this discovery as lucky, since he is not availed of all the pertinent facts. The relevant judgement for us concerns someone who (like ourselves) knows everything salient to the case, and in particular knows that Vincent is guaranteed to find this treasure. I take it that once we make clear that Vincent is guaranteed to find the treasure, however, and so form our judgement about whether the event is lucky while being fully aware of this fact, then the temptation to characterize the event as lucky disappears. The discovery is *accidental*, since Vincent wasn't aiming to find the treasure, but it is not a matter of luck that he finds treasure in this spot, as he was bound to make this discovery in this case.

Indeed, this case is in many ways akin to Rescher's example of the long-planned bequest. Although it is not an accident that the agent in Rescher's case receives the bequest (as it was planned that he should receive it), just as in Lackey's case this pleasant turn of events can seem lucky at first glance simply in virtue of it being unexpected and surprising. But once one recognizes that the target event was bound to happen then, just like Vincent's discovery of the treasure, it no longer strikes one as lucky.[28]

This brings me to the relationship between luck and control. A recurring idea in the literature on luck, particularly moral luck, is that lucky events are events that the agent lacks control over.[29] Construed as a rough necessary condition on luck, the claim is quite plausible, but so construed it is also not in any obvious tension with the modal account of luck. For if an event is within one's control to bring about, and one does bring it about, then how could it not obtain in close possible worlds where the initial conditions for that event are the same (e.g., where one continues to try to bring it about)? (Indeed, it would be a very fishy sense of "control" if it didn't generate this consequence.) It's unsurprising, then, that events that are lucky on the modal account also tend to be outside one's control. Hence there's no need for the proponent of the modal account to object to the idea that lucky events are events that are not in the agent's control (at least in some suitable sense of "control").[30]

---

[28] The similarities between this case and Rescher's example could explain why we might be tempted to characterize this example as one of good fortune. One point to keep in mind here is that Vincent is in this case finding a fortune (in treasure), and that this might well constitute background noise that impairs our judgement about the case. Would we describe a parallel case where the accidental discovery is not treasure but, say, a long-lost keepsake of sentimental rather than financial value in terms of good fortune? My instinct is that we wouldn't, and I think this reflects the fact that good fortune tends to relate to long-term significant features of one's life. For further discussion of Lackey's critique of the modal account of luck, see Levy 2009 and 2011, chap. 2. See also my footnote 23.

[29] The locus classicus when it comes to the debate about moral luck is the exchange between Nagel (1976) and Williams (1976). I offer my own response to this problem in Pritchard 2006; cf. Pritchard 2005, chap. 10. See also Driver (2012), who discusses the modal account of luck in the context of the problem of moral luck. For defences of (versions of) the lack of control account of luck, see Nagel 1976, Statman 1991, Zimmerman 1993, Greco 1995, Riggs 2007 and 2009, and Coffman 2007 and 2009. See also my footnotes 30 and 33.

[30] That said, the idea that lack of control is even a necessary condition on luck has been criticized. See, especially, Lackey 2008. For discussion, see Coffman 2009 and Levy 2009 and

The idea that lack of control is a sufficient condition for a lucky event is, however, highly dubious. To take a familiar example from the literature, the sun rose this morning, but that does not make it lucky that the sun rose—indeed, it was inevitable that it would rise.[31] It is thus incumbent upon proponents of the view that lack of control is sufficient for luck to propose a more nuanced rendering of this idea.

Here, for example, is Wayne Riggs's statement of this view (where "E" stands for the target event):

E is lucky for S iff;
(a) E is (too far) out of S's control, and
(b) S did not successfully exploit E for some purpose, and
(c) E is significant to S (or would be significant, were S to be availed of the relevant facts). (Riggs 2009, 220)

Before we start to unpack this account of luck, we should note from the outset two features of it that are controversial given our previous discussion.

The first point to note is that, following earlier work by me, Riggs opts for a significance condition on lucky events. As I explained above, however, it now strikes me as mistaken to include such a subjective factor in one's account of luck. Remember that our interest is in what makes an event lucky, and not merely on what prompts subjects to judge that an event is lucky (even though the latter can obviously be a guide to the former). With this in mind, I propose that we set condition (c) in Riggs's account of luck to one side.

The second point is related to the first, and concerns the fact that Riggs is defining not lucky events per se but rather the different notion of events that are lucky *for a subject*. With the significance condition included in the account of luck this might well make sense, in that the relevant notion of significance will probably be an agent-relative one, and hence the resulting account of luck will be agent-relative too. But in so far as we reject this condition on lucky events, is there any reason to treat an account of luck as being relativized to agents in this way? I think not, though, as we will see,

2011, chap. 2. Note that Levy 2011, chap. 2, offers a hybrid account of luck that has both a modal and a lack of control element. Given that the modal account of luck offered here is consistent with the idea that lack of control is a necessary condition for a lucky event, it is thus not obviously inconsistent with Levy's proposal. Interestingly, Levy also thinks that there is a second kind of luck besides the type that we are discussing here which concerns lucky events. This is roughly equivalent to what Nagel (1976) had in mind when he wrote about "constitutive luck," which is luck in the traits and dispositions that one has. I must confess that I am sceptical that this is a genuine kind of luck, and the reason why should be apparent from my earlier discussion—*viz.*, I think the notion that Levy has in mind is probably best understood in terms of the notion of fortune.

[31] I believe Latus (2000, 167) was the first to offer this example. See also Pritchard 2005, chap. 5.

it may well be crucial to Riggs's account that he continues to conceive of luck in this fashion.

With these two caveats in mind, let us turn our attention to conditions (a) and (b) in Riggs's account of luck. Whereas (a) is relatively clear, (b) is more opaque. In particular, what does it mean to "exploit" an event for a purpose? We can get a handle on what Riggs has in mind in this regard by considering an example that he uses (in Riggs 2009, §5).

Recall that I noted a moment ago that an obvious problem facing lack of control views concerns events like the rising of the sun which are completely out of anyone's control but which are not classed as lucky. Riggs claims that our verdict in this regard is too quick. He asks us to imagine a case where two explorers—called Smith and Jones—are about to be executed by a local tribe, only for a total eclipse to come along and for it to lead to the tribesmen abandoning their plan to kill the explorers. Riggs now imagines that while one of the explorers—Smith—had no inkling that the eclipse was going to happen, the other explorer—Jones— knew full well that this event would occur. In particular, Jones was counting on the eclipse occurring as a means of avoiding possible execution by the tribesmen.

Riggs concludes that while the event was not lucky for Jones, it was lucky for Smith. And note that this is so even despite the fact that an eclipse is a nomically necessary event that is beyond anyone's control just as much as the sun's rising in the morning. In terms of the account of luck that Riggs offers, the difference between Smith and Jones relates to condition (b). For while—in line with (a)—the event in question is equally completely beyond the control of either of them, it is only Smith who didn't successfully exploit this event for his purposes (because he didn't know it was going to occur).

That Riggs is here talking about an event that is lucky for one agent but not for another should give us pause for further reflection. For while it is undeniable that the event will seem lucky to Smith, since he was lacking crucial information about this event, I have already noted that we should not conclude from the mere fact that an event seems lucky to a certain individual that therefore it is lucky (still less that it is lucky *for that person*). Indeed, once we set aside purely subjective factors, such as one person's limited informational state, there seems no obvious reason why we would regard the eclipse happening when it did as being lucky at all—after all, it was *bound* to happen when it did. The relevant judgement to follow in this regard is thus not Smith's but ours, which (since we are in full possession of the facts) would surely accord with Jones's judgement that this event wasn't lucky at all. Furthermore, notice that once we know the facts of the situation then it matters not one jot whether we failed to "successfully exploit" the event in question, since even when it is stipulated that condition (b) is met we nevertheless do not regard the event as lucky.[32]

---

[32] Furthermore, notice that as Riggs describes the case it is also not a matter of luck that the explorers were due to be executed at that particular time. Had this not been so, then there

In any case, this is all by the by since Riggs's account still fails to explain why paradigm cases of events that are outside an agent's control don't thereby count as lucky. At best, with the example of the eclipse Riggs is offering us an example of an event which is out of the control of the agent and which doesn't satisfy the modal account of luck, but which is nonetheless (he claims) lucky for a certain agent. I have disputed this contention. The interesting question, however, is what Riggs would say about the standard case of the sun rising in the morning. Doesn't this event satisfy the conditions on luck that Riggs lays down and therefore count as lucky? Indeed, it is hard to see how adding the condition regarding the subject's failure to exploit this event for some purpose makes any difference to this perennial problem for the lack of control account of luck, given that it is *normally* the case that nomically necessary events such as this are not exploited in this way. It follows that the account of luck that Riggs is offering is untenable, at least unless one wishes to treat whole swathes of nomically necessary events as lucky. Even on a more nuanced reading, then, the lack of control account of luck is still implausible.[33]

6. Concluding Remarks

I conclude that the modal account of luck, at least when properly formulated, is still the best theory of luck available. In particular, it does not succumb to the counterexamples that have been levelled against it, and it is superior to rival proposals in the literature, such as that lucky events are events that cannot be rationally expected to occur, or that lucky events are events that are outside one's control.

Acknowledgments

My thinking about the philosophy of luck has benefitted from the input of many people over the years, including (this is *not* an exhaustive list): Nathan Ballantyne, Heather Batally, Martijn Blaauw, Kelly Becker, E. J.

would have been scope for it to be lucky that the explorers were due to be executed at the particular time in question (i.e., the time that the eclipse occurred), but Riggs also rules this out.

[33] Hurley (2003) offers what she calls a "thin" account of luck as nothing more than the obverse of responsibility, and obviously this way of thinking about luck has some parallels with the lack of control view. Given that Hurley isn't aiming to offer a full account of luck, however, I will set this proposal to one side. (Indeed, on the face of it, at least, it seems that what Hurley has in mind could be captured by the idea, compatible with the modal account of luck, that lack of control is a necessary condition for lucky events.) Similar points apply to Mele's (2006) conception of luck. Like Hurley, Mele doesn't seem to be offering a complete account of luck (or, at least, I have struggled to discern one from the text), though he does appear to hold that lack of control is at least a necessary condition for lucky events. So construed, however, his conception of luck will also be compatible with the modal account of luck offered here.

Coffman, Steven Hales, Jennifer Lackey, Joe Milburn, Nicholas Rescher, Wayne Riggs, Sabine Roeser, Matthew Smith, Lee John Whittington, and Di Yang.

References

Ballantyne, N. 2011. "Anti-Luck Epistemology, Pragmatic Encroachment, and True Belief." *Canadian Journal of Philosophy* 41:485–504.

———. 2012. "Luck and Interests." *Synthese* 185:319–34.

Becker, K. 2007. *Epistemology Modalized*. London: Routledge.

Black, T. 2002. "A Moorean Response to Brain-in-a-Vat Scepticism." *Australasian Journal of Philosophy* 80:148–63.

———. 2008. "Defending a Sensitive Neo-Moorean Invariantism." In *New Waves in Epistemology*, ed. V. F. Hendricks and D. H. Pritchard, 8–27. Basingstoke: Palgrave Macmillan.

———. 2011. "Modal and Anti-Luck Epistemology." *Routledge Companion to Epistemology*, ed. S. Bernecker and D. H. Pritchard, 187–98. London: Routledge.

Black, T., and P. Murphy. 2007. "In Defense of Sensitivity." *Synthese* 154:53–71.

Coffman, E. J. 2007. "Thinking About Luck." *Synthese* 158:385–98.

———. 2009. "Does Luck Exclude Control?" *Australasian Journal of Philosophy* 87:499–504.

Domsky, D. 2004. "There Is No Door: Finally Solving the Problem of Moral Luck." *Journal of Philosophy* 101:445–64.

Dretske, F. 1970. "Epistemic Operators." *Journal of Philosophy* 67:1007–23.

———. 1971. "Conclusive Reasons." *Australasian Journal of Philosophy* 49:1–22.

Driver, J. 2012. "Luck and Fortune in Moral Evaluation." In *Contrastivism in Philosophy*, ed. M. Blaauw, 154–72. London: Routledge.

Enoch, D., and E. Guttel. 2010. "Cognitive Biases and Moral Luck." *Journal of Moral Philosophy* 7:372–86.

Fantl, J., and M. McGrath. 2007. "On Pragmatic Encroachment in Epistemology." *Philosophy and Phenomenological Research* 75:558–89.

———. 2011. "Pragmatic Encroachment." *Routledge Companion to Epistemology*, ed. S. Bernecker and D. H. Pritchard, 558–68. London: Routledge.

Greco, J. 1995. "A Second Paradox Concerning Responsibility and Luck." *Metaphilosophy* 26:81–96.

———. 2007. "Worries About Pritchard's Safety." *Synthese* 158:299–302.

Hawthorne, J. 2004. *Knowledge and Lotteries*. Oxford: Clarendon Press.

Hurley, S. 2003. *Justice, Luck, and Knowledge*. Cambridge, Mass.: Harvard University Press.

Kahneman, D., and C. A. Varey. 1990. "Propensities and Counterfactuals: The Loser That Almost Won." *Journal of Personality and Social Psychology* 59:1101–10.

Kruger, J., and D. Dunning. 1999. "Unskilled and Unaware of It: How Difficulties in Recognizing One's Own Incompetence Lead to Inflated Self-Assessments." *Journal of Personality and Social Psychology* 77:1121–34.

Lackey, J. 2008. "What Luck Is Not." *Australasian Journal of Philosophy* 86:255–67.

Langer, E. J. 1975. "The Illusion of Control." *Journal of Personality and Social Psychology* 32:311–28.

Latus, A. 2000. "Moral and Epistemic Luck." *Journal of Philosophical Research* 25:149–72.

Levy, N. 2009. "What, and Where, Luck Is: A Response to Jennifer Lackey." *Australasian Journal of Philosophy* 87:489–97.

———. 2011. *Hard Luck: How Luck Undermines Free Will and Moral Responsibility*. Oxford: Oxford University Press.

Lewis, D. 1973. *Counterfactuals*. Oxford: Blackwell.

———. 1987. *On the Plurality of Worlds*. Oxford: Blackwell.

Luper, S. 1984. "The Epistemic Predicament." *Australasian Journal of Philosophy* 62:26–50.

———. 2003. "Indiscernability Skepticism." In *The Skeptics: Contemporary Essays*, ed. S. Luper, 183–202. Aldershot: Ashgate.

McKenna, F. P. 1993. "It Won't Happen To Me: Unrealistic Optimism or Illusion of Control?" *British Journal of Psychology* 84:39–50.

Mele, A. 2006. *Free Will and Luck*. Oxford: Oxford University Press.

Nagel, T. 1976. "Moral Luck." *Proceedings of the Aristotelian Society* 50 suppl.: 137–52.

Nozick, R. 1981. *Philosophical Explanations*, Oxford: Oxford University Press.

Pritchard, D. H. 2002. "Resurrecting the Moorean Response to the Sceptic." *International Journal of Philosophical Studies* 10:283–307.

———. 2004. "Epistemic Luck." *Journal of Philosophical Research* 29:193–222.

———. 2005. *Epistemic Luck*. Oxford: Oxford University Press.

———. 2006. "Moral and Epistemic Luck." *Metaphilosophy* 37:1–25.

———. 2007a. "Anti-Luck Epistemology." *Synthese* 158:277–97.

———. 2007b. "Knowledge, Luck, and Lotteries." In *New Waves in Epistemology*, ed. D. H. Pritchard and V. Hendricks, 28–51. London: Palgrave Macmillan.

———. 2008. "Sensitivity, Safety, and Anti-Luck Epistemology." In *Oxford Handbook of Scepticism*, ed. J. Greco, 437–55. Oxford: Oxford University Press.

———. 2009. "Safety-Based Epistemology: Whither Now?" *Journal of Philosophical Research* 34:33–45.

———. 2012a. "Anti-Luck Virtue Epistemology." *Journal of Philosophy* 109:247–79.

———. 2012b. "In Defence of Modest Anti-Luck Epistemology." In *The Sensitivity Principle in Epistemology*, ed. T. Black and K. Becker, 173–92. Cambridge: Cambridge University Press.

———. 2012c. "The Methodology of Epistemology." *Harvard Review of Philosophy* 18:91–108.

———. 2013. "There Cannot Be Lucky Knowledge." In *Contemporary Debates in Epistemology*, 2nd ed., ed. M. Steup, J. Turri, and E. Sosa, 152–64. Oxford: Blackwell.

———. 2014a. "Risk." Unpublished manuscript.

———. 2014b. "Sceptical Intuitions." In *Intuitions*, ed. D. Rowbottom and T. Booth, 213–31. Oxford: Oxford University Press.

———. Forthcoming *a*. "Anti-Luck Epistemology and the Gettier Problem." *Philosophical Studies*.

———. Forthcoming *b*. "Engel on Pragmatic Encroachment and Epistemic Value." *Synthese*.

———. Forthcoming *c*. "Knowledge, Luck and Virtue: Resolving the Gettier Problem." In *The Gettier Problem*, ed. C. Almeida, P. Klein, and R. Borges. Oxford: Oxford University Press.

Pritchard, D. H, and M. Smith. 2004. "The Psychology and Philosophy of Luck." *New Ideas in Psychology* 22:1–28.

Rescher, N. 1990. "Luck." *Proceedings and Addresses of the American Philosophical Association* 64:5–19.

———. 1995. *Luck: The Brilliant Randomness of Everyday Life*. New York: Farrar, Straus and Giroux.

Riggs, W. 2007. "Why Epistemologists Are So Down on Their Luck." *Synthese* 158:329–44.

———. 2009. "Luck, Knowledge, and Control." In *Epistemic Value*, ed. A. Haddock, A. Millar, and D. H. Pritchard, 205–21. Oxford: Oxford University Press.

Roush, S. 2005. *Tracking Truth: Knowledge, Evidence and Science*. Oxford: Oxford University Press.

Royzman, E., and R. Kumar. 2004. "Is Consequential Luck Morally Inconsequential? Empirical Psychology and the Reassessment of Moral Luck." *Ratio* 17:329–44.

Sainsbury, R. M. 1997. "Easy Possibilities." *Philosophy and Phenomenological Research* 57:907–19.

Sosa, E. 1999. "How to Defeat Opposition to Moore." *Philosophical Perspectives* 13:141–54.

Statman, D. 1991. "Moral and Epistemic Luck." *Ratio* 4:146–56.

Steglich-Peterson, A. 2010. "Luck as an Epistemic Notion." *Synthese* 176:361–77.

Svenson, O. 1981. "Are We Less Risky and More Skillful Than Our Fellow Drivers?" *Acta Psychologica* 47:143–51.

Teigen, K. H. 1995. "How Good Is Good Luck? The Role of Counterfactual Thinking in the Perception of Lucky and Unlucky Events." *European Journal of Social Psychology* 25:281–302.

------. 1996. "Luck: The Art of a Near Miss." *Scandinavian Journal of Psychology* 37:156–71.

------. 1997. "Luck, Envy, Gratitude: It Could Have Been Different." *Scandinavian Journal of Psychology* 38:318–23.

------. 1998a. "Hazards Mean Luck: Counterfactual Thinking and Perceptions of Good and Bad Fortune in Reports of Dangerous Situations and Careless Behaviour." *Scandinavian Journal of Psychology* 39:235–48.

------. 1998b. "When the Unreal Is More Likely Than the Real: *Post Hoc* Probability Judgements and Counterfactual Closeness." *Thinking and Reasoning* 4:147–77.

------. 2003. "When a Small Difference Makes a Large Difference: Counterfactual Thinking and Luck." In *The Psychology of Counter-factual Thinking*, ed. D. R. Mandel, D. Hilton, and P. Catellani. London: Routledge.

Tetlock, P. E. 1998. "Close-Call Counterfactuals and Belief-System Defenses: I Was Not Almost Wrong but I Was Almost Right." *Journal of Personality and Social Psychology* 75:639–52.

Tetlock, P. E., and R. N. Lebow. 2001. "Poking Counterfactual Holes in Covering Laws: Cognitive Styles and Historical Reasoning." *American Political Science Review* 95:829–43.

Thompson, S. C. 1999. "Illusions of Control: How We Overestimate Our Personal Influence." *Current Directions in Psychological Science* 8:187–190.

------. 2004. "Illusions of Control." In *Cognitive Illusions: A Handbook on Fallacies and Biases in Thinking, Judgement and Memory*, ed. R. F. Pohl, 115–25. Hove, U.K.: Psychology Press.

Turri, J., and O. Friedman. 2014. "Winners and Losers in the Folk Epistemology of Lotteries." In *Advances in Experimental Epistemology*, ed. J. Beebe, chap. 6. London: Bloomsbury.

Unger, P. 1968. "An Analysis of Factual Knowledge." *Journal of Philosophy* 65:157–70.

Wagenaar, W. A. 1988. *Paradoxes of Gambling Behaviour*. Hove, U.K.: Lawrence Erlbaum.

Wagenaar, W. A., and G. B. Keren. 1988. "Chance and Luck Are Not the Same." *Journal of Behavioural Decision Making* 1:65–75.

Williams, B. 1976. "Moral Luck." *Proceedings of the Aristotelian Society* 50 suppl.: 115–35.

Williamson, T. 2000. *Knowledge and Its Limits*. Oxford: Oxford University Press.

Zimmerman, M. 1993. "Luck and Moral Responsibility." In *Moral Luck*, ed. D. Statman, 217–34. Albany: State University of New York Press.

CHAPTER 9

THE MACHINATIONS OF LUCK

NICHOLAS RESCHER

1. Luck's Partners: Fate and Fortune

Luck would appear to be one of the prime potencies that afflict the condition of people for better or for worse. Yet perhaps the single most difficult thing in the theory of the subject is to say just exactly what "luck" is so as to enable us to distinguish its operation from that of other factors at work in human affairs.

The evaluation of what happens is pivotal for luck. When a condition or an occurrence impacts positively or negatively upon someone, this happens in general through one of three agencies:

- *Nature*: the world's impersonal arrangements and occurrences, the course of developments in the world at large.
- *Effort*: the actions and activities that people deploy to modify and shape the conditions of their lives. (A person has talents and abilities through impersonal nature but develops and applies them by thought-directed effort.)
- *Chance*: the unforeseen, unpredictable, and unplanned occurrences that affect someone for good or ill.

Corresponding—perhaps somewhat loosely—to this trichotomy are three potencies at work in shaping the condition of people, namely:

- *Fate*, which is a matter of what nature does to and for us. It is by fate that you are handsome and smart—or the reverse.

The Philosophy of Luck, First Edition. Edited by Duncan Pritchard and Lee John Whittington.
Chapters © 2014 Metaphilosophy LLC and John Wiley & Sons Ltd, except for "Luck as Risk and the Lack of Control Account of Luck" © 2015 Metaphilosophy LLC and John Wiley & Sons Ltd. Book compilation © 2015 Metaphilosophy LLC and John Wiley & Sons Ltd.
Published 2015 by John Wiley & Sons Ltd.

- *Fortune*, which is a matter of what we ourselves make of the opportunities at our disposal. (*Faber est suae quisque fortunae*, as the Roman maxim had it.)
- *Luck*, which is a matter of those goods and bads that befall us purely by chance, in a way that is unforeseen, unplanned for, and unexpected—at any rate by the agent herself.

Luck is at issue whenever it is a matter of pure chance that a result of significant positive or negative value is realized for someone: good luck when the outcome is positive and bad luck when it is negative.

The idea of luck revolves about the operation of these varying potencies in the area of human affairs. It is by fate that you are rich if you inherit your millions, by fortune if you acquire them by hard work and business acumen, and by luck if you win the lottery. Of the five youngsters who took an exam, only Tommy passed it. Was it lucky he did so? Presumably not. He was an able student and came prepared. His passing the exam was a fortunate development, but presumably he did not do so by luck.

With luck pure chance is critically involved. No doubt nothing whatever is unplanned by and unforeseeable to a God who tracks more than the flight of every sparrow. But there is much that lies outside the ken of us imperfect and ignorant humans.

Whether we are born in Julius Caesar's day or the twenty-first century is a matter of fate, as is the issue of whether our parents are paupers or billionaires. (It is not *a matter of luck*—one is not there, preexistingly, so that chance can do one favors.)

Marrying the daughter of a Vanderbilt or a Rockefeller is a matter of fortune; one has to woo and win the girl by means of charm and perseverance. (It is *not a matter of luck*—it is by exerting persistence and effort in the pursuit of a plan.) (The situation would of course be different—and indeed lucky—if a shipwreck stranded the two of you as sole occupants of some desert isle.)

A card player plays the wrong card. This could have occurred because he didn't know any better—that's unfortunate. But it could also be that he has accidentally taken the wrong one—that's unlucky. What he has performatively done is the same thing either way. But the way in which it has come about—its explanatory rationale—makes all the difference.

It is bad enough that fate, fortune, and luck are sufficiently related that people all too commonly confuse them, but their interconnection in life's affairs can be hard to disentangle.

Robinson's car was hit at a level crossing just as the 9:45 A.M. train was coming by. He was certainly unlucky in that his car stalled at that particular time. And he was doubly unlucky because this train was almost always delayed. But one could also say that he is unfortunate in not having had the vehicle properly serviced. That particular requirement for the mishap was entirely his own doing. In such cases one has to be specific as to the realization at issue. We have to say not just that Robinson was unlucky but that he was unlucky in that his car stalled out at the time, or in that the 9:45 was on time. And we can also say that he was unfortunate in not having had his car properly maintained. Both luck and fortune collaborated in producing Robinson's disaster.

Smith was a hotly contested electee as mayor. Which potency gets the credit? Possibly all three are involved. Smith's inborn talents as administrator and persuasive speaker may have positioned him to enter the race. (So, think fate.) His hard work in networking and fundraising may have made him a prime contender. (So, credit fortune.) But Smith won by only five votes in his neck-to-neck rivalry with Jones, twenty of whose supporters were on a bus that broke down in a last-minute rush to beat the poll-closing time. (That was pure luck.) All three contributions were essential to the result. And things often work like that in life's affairs. Those three life-shaping potencies will generally interact, working with or against one another in the production of outcomes.

Sometimes it is hard to say whether nature or pure chance is in control. Consider the general attending Hitler's "Wolf's Lair" conference on the fateful morning of the assassination attempt who left the conference room on a "call of nature" just before the bomb exploded, and so survived unharmed. Would we say that he was lucky on this occasion or merely fortunate? It is hard to know just where to draw the line.

The night before the mayoral election, all of Smith's—decidedly abler—competitors were laid low by a stomach virus and thus won by default—and against all reasonable expectation. Was Smith lucky to win? The answer will depend on the reference points of judgment. At the start of the race Smith's winning was effectively a foregone conclusion. There was no way he could lose. But a week beforehand matters looked quite different. Here the odds looked to be very much against Smith, and his winning looked to be very unlikely. On this perspective he turned out to be lucky.

2. Luck Is Statistical in Its Dependence on the Prevailing Context

Being probabilistic in nature, luck is a matter of the composition of reference classes. Consider a ship that founders in a storm at sea. It is carrying a hundred passengers, fifty men as well as fifty women and children. Forty men drown, but—rather gallantly—none of the women and children. An otherwise unspecified passenger has a 60 percent chance of survival, and would be somewhat lucky to be rescued. But only 20 percent of the men are saved, so a rescued man is very lucky indeed. With the women and children, being saved looks to be a certainty, and chance does not seem to come into it. They are fortunate to be saved, but not lucky. With luck chance is pivotal, and the comparison group that serves as the frame of reference is thus pivotal for luck.

The idea of good (or bad) luck is inherently context relative: one is not simply lucky with regard to the outcome as such but rather lucky in relation to the circumstances that surround it. The lottery winner is lucky seeing that so many other tickets were sold. The person who survives a tsunami is lucky in that so many others did not. The circumstances as well as the outcome are crucial factors here.

3. Is Luck Objective or Subjective?

Luck will be either objective or subjective, depending upon whether the outcome at issue *is in fact* beneficial or harmful, or whether it *is merely viewed* in a positive or a negative light.

Subjective luck depends on the expectations of the individual; objective luck hinges on the actualities of the situation. If, unbeknownst to me, some-one has fixed the lottery in my favor, I may be fortunate to win, but no longer objectively lucky. And in cases where my disappointed expectation is unrealizably high, I will deem myself more unlucky than I actually was. In the light of such considerations there will have to be due heed of the difference between actual and apparent luck.

4. Luck Depends on What Follows

With luck the course of ultimate development becomes critical, and the issue of eventual outcome is a salient consideration. With four equally plau-sible rivals pressing in, Henry was lucky to have won the hand of the rich widow Jones. But his good luck would no longer continue to count as such if it turns out she is a nasty shrew who makes the life of her husband miser-able. And someone who survives a disaster by good luck may no longer see matters in this light if this survival does no more than set her up for some sort of horrendous catastrophe.

In general, however, we talk of luck in relation to proximate rather than ultimate outcomes, the latter being—often as not—altogether unforeseeable.

5. Can One Control Luck?

Can luck be brought under control? Can one make one's own luck?

In relation to this issue a careful distinction must be drawn. Luck (by definition) is a matter of eventuations due to pure chance, and chance (by definition) is something that is not in one's control. What one certainly can do is to open oneself up to the intervention of luck—to act so as to give chance a greater or lesser prospect for impacting one's affairs. Avoiding bad luck calls for prudence and risk aversion: acting with caution, buying insur-ance, and the like. Good luck, by contrast, can be sought via crucially cal-culated risks. The person who does not buy a ticket cannot win the lottery.

Only in running risks do we open ourselves to the prospect of good luck. And we live in a world where risks and benefits are often inversely correlated with one another. Only the Stoic who schools himself to be indifferent to the occurrences of life immunizes himself against the impact of luck.

6. Can One Measure Luck?

People can of course be more lucky or less lucky. The person who is very lucky is one who realizes a great benefit against the odds; and the person who is very unlucky is one who does the reverse. One key factor in determining luck is the difference between the actual outcome and what might have been: if the outcome is favorable, the agent is lucky to the extent that this result differs from its unfavorable alternative; if the outcome is unfavorable, the agent is unlucky to the extent that this differs from what would have been if things had gone well. Either way, the difference in value between an unfavorable and a favorable outcome is crucial for the extent of luck.

The second key determinant of luck is probability. An agent is the more lucky not only with a favorable result that is of greater value but also with one that is more unlikely.

On this basis, let Δ represent the value difference between a favorable result and an unfavorable one—that is, the value that is at stake. And let p be the probability of success (that is, of a favorable result), so that $1 - p$ measures the probability of failure (that is, of a negative result). Then the following measure of the amount of luck λ would be a good first attempt:

(1) Favorable result

$$\lambda^+ = \Delta \times (1 - p) = \Delta - \Delta p$$

(2) Unfavorable result

$$\lambda^- = -\Delta p$$

So when one views the failure probability $(1 - p)$ as a measure of the *risk*, and the difference between a favorable and an unfavorable outcome (Δ) as a measure of the *stake*, then the amount of (good) luck at issue with a favorable result is simply the product of these two quantities: risk × stake.[1] On this basis, winning $500 at a 90 percent favorable chance (λ value of $(1 - \frac{9}{10}) \times 500 = 50$) is equivalent with winning $ 100 at an even chance (a merely 50:50 favorable chance: $(1 - \frac{1}{2}) \times 100 = 50$).

---

[1] Note that the difference between good luck (λ^+) and bad luck (λ^-) is exactly the stake at issue: $\lambda^+ - \lambda^- = \Delta$.

An instructive way of looking at the matter is to regard the situation as a gamble with Δ at stake and p the probability of winning it. Then the value of bad luck is simply the loss (negative) of one's expectation—that is, $-p\Delta$—and that of good luck is simply the gained stake itself discounted by the expectation of obtaining it, $\Delta - p\Delta$, that is, the excess of the realized result over what one is entitled to expect.

7. Retrospect

The interplay of fate, fortune, and luck in human life is seldom stressed sufficiently by biographers, who often depict the life story of their protagonists as the inevitable unfolding of an unbreakable chain of circumstance. The reality of it is generally quite different. Nature emplaces us in a setting of place, time, and situation of its own choosing. The inner impetus of character and personality leads us to develop or neglect our talents and capacities. Pure chance brings our way certain opportunities and prospects which we may or may not seize. One darn thing leads to another—much or most of them outside our control. Constructing a coherent life place out of the flotsam and jetsam of possibility and opportunity is a task well beyond the capacity of most of us, who generally simply "go with the flow," but when it happens luck generally discovers a lion's share of the credit.

8. Concluding Worries

But a big question lurks in the background. Even if one grants that the preceding analysis adequately characterizes how people *think* of luck, the question yet remains: Is there really any such thing?

There are certainly doctrinal positions that deny the existence of luck. In the main they are of three types:

- *Mechanistic determinism*: The world is one vast machine of sorts, all of whose operations are unavoidingly predetermined by nature's inexorable laws. Determinative atomists like Lucretius, mechanistic determinists like Laplace, and dialectical materialists like Marx and Engels have held positions of this sort.
- *Metaphysical determinism*: The world's eventuations are one and all predetermined for the very interest of time by principles of lawful order that necessitates all of its occurrences. The ancient Stoics, Spinoza, and scientistic positivists like Comte maintained this sort of thing.
- *Theological predestinationism*: Some have envisioned the world's treaty as the temporal unfolding of a vast and all-determinative program through which God sets into action an all-determinative plan by which all of history's developments are predetermined with inexorable certainly like the events unfolding when a film-strip is played. Islamic fatalists and Calvinists are of this persuasion.

None of these positions allows any room for chance, accident, and choice contingency in the world's scheme of things. As their proponents see it, the idea of objective luck is an illegitimate illusion. At most and at best there is room only for subjective luck rooted in humankind's incomplete and imperfect knowledge of how things happen in the world. Luck is no more than a misimpression rooted in human cognitive imperfection and ignorance.

In the wake of the prominent role allotted by modern science to choice and chance in the world's scheme of things—the stochastic operation of physical nature and the drastic complexities of brain processes—such a position is difficult to maintain. In the world order replete with probabilistic principles, denying the role of scheme and haphazard is no easy prospect.

Perhaps a day will come when the idea of luck can be abandoned, with chance and contingency dismissed as a factor in human affairs and Bishop Butler's dictum that probability is the guide of life no longer in order. But from the vantage point of present indications, such a prospect looks to be extremely remote.[2]

The condition of humankind being as it to all appearances is in this uncertain world, luck cannot be eliminated as a key factor of our existence, be it in cognitive, practical, or ethical contexts.

Reference

Rescher, Nicholas. 1995. *Luck*. New York: Farrar, Straus and Giroux (reprinted Pittsburgh: University of Pittsburgh Press, 2002).

---

[2] This discussion builds up and develops some ideas in Rescher 1995.

CHAPTER 10

LUCK, KNOWLEDGE, AND "MERE" COINCIDENCE

WAYNE D. RIGGS

1. Introduction

There is a thriving cottage industry within the theory of knowledge that approaches the problem of analyzing knowledge by making appeal to the notion of luck. Mere appeals to luck have been common in epistemological discussions for a long time, but only recently has the thought been pursued in a thoroughgoing way that one might usefully make better sense of what *knowledge* amounts to by thinking hard about what *luck* amounts to. The move is a tempting one for several reasons.

First, it seems quite natural to characterize what goes wrong in most cases of true belief that do not amount to knowledge in terms of luck. A paradigm case of true belief that doesn't rise to the status of knowledge is the "lucky guess." Gettier cases have been plausibly described as always including an element of "double luck" (see Zagzebski 1999). In general, there has been widespread agreement that knowledge can't be a matter of luck. There is no such thing as "lucky knowledge." There have been dissenters, of course,[1] but if there is a consensus view in epistemology, this is surely it.

Second, at least some luck theories of knowledge allow for a kind of conceptual unification between epistemology and other subdisciplines within philosophy. For example, if one thinks of knowledge as a kind of achievement (or success through ability [see Greco 2010]), then any instance of knowing is simply a token of this broader type of achievement. Hence, whatever philosophical account we give of achievements will serve as an

---

[1] E.g., Stephen Hetherington.

The Philosophy of Luck, First Edition. Edited by Duncan Pritchard and Lee John Whittington. Chapters © 2014 Metaphilosophy LLC and John Wiley & Sons Ltd, except for "Luck as Risk and the Lack of Control Account of Luck" © 2015 Metaphilosophy LLC and John Wiley & Sons Ltd. Book compilation © 2015 Metaphilosophy LLC and John Wiley & Sons Ltd. Published 2015 by John Wiley & Sons Ltd.

account of knowledge as well. And luck is quite plausibly of central importance to the concept of achievement (or success through ability). This kind of unification is often considered a theoretical virtue.

Third, this general approach to the theory of knowledge is more top-down than most. That is, it does not begin with a painstaking catalogue of particular scenarios and cases, building a defense of the view by its successful treatment of those cases. This is a standard approach in contemporary epistemology, and I do not mean to impugn it. However, after many years of subjecting every extant theory of knowledge to an exhausting barrage of counterexamples, it is hard to make further headway following this procedure. But if we can give a convincing argument that knowledge can be analyzed primarily in terms of luck, then we gain new conceptual and dialectical resources. Of course, any theory of knowledge will eventually be accountable to the standard counterexamples. But when substantial support for the theory comes via another route—for example, the plausibility of the conceptual connection between luck and knowledge—isolated failures to give the "right" answer in particular cases can be overlooked with less fear of the charge of ad hocness.

Fourth, and related to the last point, a theory of knowledge that appeals centrally to the notion of luck gets to avail itself of all our standard intuitions about luck, both in general and in particular cases. Granted this affords additional opportunities for one's theory to fail to tally with those intuitions, but it also provides a source of support that is somewhat independent of our usual intuitions regarding knowledge. If we could analyze knowledge in terms of luck, we would gain a new window on the subject matter of epistemology. We could turn to platitudes about luck to guide us in our theorizing about knowledge. Our intuitions about luck would turn out to be intuitions about knowledge as well. Since philosophers' intuitions about knowledge are well plumbed by this point, a new set of intuitions that are relatively independent of the standard positions and arguments in epistemology would be a breath of fresh rhetorical air. And if one can develop a theory of knowledge that does reasonable justice to our intuitions about both knowledge and luck, one has got a powerful view.

Of course, if a theory of knowledge put in terms of luck is to move forward our understanding, we need to have an account of luck as well. In my own work (e.g., Riggs 2007, 2009a, and 2009b), I have defended a theory of knowledge according to which S knows that p so long as S's believing the truth about p is not a matter of luck. Luck, in turn, was defined in terms of the extent to which an agent has control over an outcome. Hence, S knows that p if and only if S's believing the truth about p is an outcome that is/was sufficiently under S's control. This kind of view is sometimes referred to as a "control theory" of luck, a term that I will adhere to in this essay.

For reasons I have developed elsewhere, I think the control theory provides a very promising general account of luck. There are, however, both problems with theories of luck in general and problems specific to control theories. Regarding the former, all the above advantages of looking to a theory of luck for insight into our theory of knowledge are counterbalanced by some significant disadvantages. It turns out that giving a precise and convincing account of luck is no easier than giving an account of knowledge in the first place. Just as with analyzing knowledge, there are many overlapping and related notions in the conceptual neighborhood that muddy the waters and threaten to force anyone who hopes to make progress to make ad hoc stipulations just to get some traction on the problem. Providing a fully general account of luck is a worthy project in its own right, but it is not necessarily a prudent strategy for pursuing an account of knowledge.

There are many objections that apply specifically to control theories of luck as well. My purpose here is not to defend the control theory against any of these but rather to develop in section 3 one particularly troubling objection and some possible responses to it. In the end, I'm not sure the responses are convincing, but they do suggest a slightly different successor theory of knowledge that does not appeal to such a general theory of luck. The final section of the essay briefly develops this idea in hopes of keeping many of the benefits of luck theories of knowledge while avoiding some of the pitfalls. But first I will catalogue what I take to be some of the more significant frustrations involved in providing a general theory of luck.

2. Having No Luck with "Luck"

There is now a respectable and growing literature on the topic of the nature of luck itself. There is a variety of different kinds of theories of luck—the control theory, the safety theory (e.g., Pritchard 2005), the probability theory (e.g., Rescher 1995), and variants of each of these. The topic is fascinating, and the literature rich and insightful. However, the field of "luckology" has problems very similar to the problems one finds in epistemology proper. The term "luck" comes with a whole host of semantic and affective associations, much like the term "knowledge." These associations are triggered whenever one is asked to judge intuitively whether some described event, whether real or imagined, is "lucky" or not.

For instance, consider the event of winning a fair lottery with one million tickets. I take this to be an absolute paradigm case of a lucky event. Notice that it has all the features highlighted by most of the various theories of luck—it is highly improbable, completely out of the control of the winner, could easily have been otherwise, is extremely significant to the winner, and so on. In this case, all (or at least most) of the things we associate with lucky events are present. Everyone can agree that this is a case of luck and that the lottery winner is thereby lucky.

But there are other cases not so clear cut. One might think that the survivor of a game of Russian Roulette is lucky to be alive, yet her survival is quite probable (assuming one bullet and, say, six chambers in the revolver). We are also likely to say the same thing about a passenger on a plane that unexpectedly came very close to colliding with another plane due to a very unlikely series of pilot and air traffic controller errors. This case seems especially odd since, by hypothesis, the close call was itself incredibly unlikely. It would seem that the passenger was, if anything, *unlucky* as a passenger on that flight. There are events that are clearly out of any human's control and yet seem obviously not lucky or unlucky for anyone, there are events that are lucky yet apparently insignificant, and so on for every one of the properties commonly associated with lucky events, all of which happily coincide in a standard lottery example. This psychological and perhaps conceptual entanglement of the disparate notions of safety, chance, probability, significance, control, and the rest makes thinking clearly about the nature of luck very difficult.

The problem doesn't end with the terms "luck" and "lucky" either. There is a whole family of terms that capture bits and pieces of this tangled web of associations: luck, fortune, accident, happenstance, chance, and the like. Yet there are differences in nuance among each of these terms, such that we are loath to treat any as quite synonymous with any other. For instance, consider E. J. Coffman's treatment of Nicholas Rescher's account of luck. Rescher claims that improbability is not a necessary condition on luck and he gives the Russian Roulette case as a demonstration of this. Coffman disagrees, claiming that the Russian Roulette survivor is actually not lucky after all. He provides an error theory to explain the appearance to the contrary in terms of the distinction between luck and fortune: "You can be fortunate with respect to an event whose occurrence was *extremely* likely, whereas an event is *lucky* for you only if there was a significant chance the event wouldn't occur. ... Now, because *luck and fortune* are closely related, we shouldn't be shocked to learn that they are sometimes confused with one another" (Coffman 2007, 392). The details of this exchange need not detain us here. The point is that our task of making good sense of the notion of luck is exacerbated by its conceptual proximity to these closely related notions.

There are also affective connotations of the term "lucky" that make matters difficult. We tend to naturally think of someone who is lucky as someone who has had something *good* happen to him. Otherwise, he would be *unlucky*. (And, of course, if something has happened to him that counts as good, it is clearly significant, bringing that notion into play as well.) I think this is, in part, what motivates Nathan Ballantyne to claim that "[s]ome lucky events involve greater luck than more unlikely lucky events" (Ballantyne 2014, 1399). He takes it to be intuitively obvious that the winner of $1 in a thousand-ticket lottery is not as lucky as the winner of $100,000 in a 995-ticket lottery. The force behind this intuition seems to be that the winner of $100,000 has just *benefited* a great deal more than the winner of $1. The event was *better* for her, and hence she is *luckier.*

And finally, there is the problem of narrowing down just what it is that we are calling "lucky" in the first place. When one gets down to cases and begins the business of either picking away at or defending a particular view of luck, the object of discussion can get very slippery. Jennifer Lackey proposes a counterexample to the control theory of luck (Lackey 2008, 258). In her example, Ramona is attempting to demolish an old warehouse. She has wired explosives to a button on her desk. Unbeknownst to her, a mouse has chewed through the wires connecting the button to the explosives, severing the connection. But just before Ramona presses the button, a co-worker hangs his jacket on a nail that happens to be in the precise location of the break. This completes the circuit, and the explosives go off. The explosion itself appears to be very lucky for Ramona, yet it also appears that Ramona simply did something within her control—she pushed a button that set off the explosives.

   In a reply to this objection, Coffman again presents an error theory to explain why it appears that Ramona was lucky the explosives went off, even though she isn't: "Just before Ramona pressed the button, she became free to do something that would definitely cause the explosion. *That* was a stroke of good luck. But Ramona's becoming free to cause the explosion differs from the explosion itself. Whoever finds it obvious that Ramona was lucky to explode the building would seem to be confusing the explosion itself with Ramona's becoming free to cause the explosion" (Coffman 2009, 502). Coffman seems right to insist that we be very clear about precisely which event is under consideration when we talk about luck. And he seems right to say that the explosion is distinct from Ramona's being free to cause the explosion. But virtually every putatively lucky (or not lucky) event will have causes that can be finely individuated in this way. Is the lottery winner lucky to have $100,000 or is she merely lucky to have had her ticket selected? Or is she lucky to have bought that ticket rather than another? In at least some cases, the luck involved at whatever stage in the causal process it occurs seems to carry through to the outcome. If we decide that the winner was lucky to have had her ticket selected, we are quite happy to say that the luck involved there gets "transmitted" through to the outcome of receiving $100,000. But Coffman argues that there are cases in which the transmission fails. So we have yet another layer of complexity to attend to in our theory of luck.

3. Sunrise Cases

In my own previous work on luck, I have defended a control analysis, both in its own right and as a component in a theory of knowledge. On my theory, an outcome e is lucky for S to the extent that the occurrence of e was out of S's control (see especially Riggs 2009a). We do not need to get into the details about what constitutes control at this point to see what I consider to be one of the major problems facing this sort of view. For convenience, let us call the control condition just elucidated the "lack of control" requirement, or LCR.[2] The problem itself is also easily stated: lots of events are outside

---

[2] I'm following Coffman 2009 loosely here.

our control yet do not intuitively count as lucky. A standard example of such an event is the "rising" of the sun each morning.[3] This event is far from my control, and yet it seems absurd to claim that I am "lucky" that the sun rose this morning or any other morning.

It seems to me that there are two fairly obvious ways for the control theorist to respond to this problem. One is to add a "significance" requirement to the theory in addition to LCR. That is, in order for an event to count as lucky for S, it must be significant to S in some way. Since the sunrise is a normal occurrence that is generally not especially significant for anyone (aesthetic sensibilities aside), such cases would not count as lucky on such an amended view. This is the tactic I chose in my previous work.

The other obvious move would be to provide an error theory in terms of Gricean implicatures. On this view, sunrises and other mundane but uncontrollable events really are lucky. Since everyone is generally aware, however, that such mundane events take place as a matter of course, and that they are not under anyone's control, it would violate the Gricean maxim of relevance to bother saying something like "We sure are lucky the sun rose this morning." After all, on this version of the control theory, almost everything that ever happens is a matter of luck for each of us. Hence, in order for it to be worth mentioning, there must be something significant about the fact that the event being mentioned is out of one's control.

I don't find either of these approaches very appealing. The first feels somewhat ad hoc. If what is really central to something's being a matter of luck for S is S's lack of control over it, then why would it matter whether or not the outcome under discussion is significant or not? The two conditions, LCR and significance, seem unrelated except insofar as our intuitions about what is "lucky" force them to coincide. There is a strong nagging suspicion that if we just got the other condition right, the solution to the significance problem would simply fall out of it. Otherwise, why would two such otherwise apparently unrelated conditions be wedded in such a fundamental and ubiquitous notion as luck?

Moreover, there are lots of outcomes that seem to be matters of luck that are *insignificant*. If I flip a fair coin and it comes up heads, that is a matter of luck even if nothing hangs on it. Adding a significance condition to solve the problem of "sunrise cases" for the control theory seems to be overkill if it forces us to say that *every* event that is a matter of luck must be of some significance.[4]

The second approach ultimately seems somewhat more appealing. It allows one's actual theory of luck to be less complicated—one would require only the LCR to capture the notion. Every event that is out of our

---

[3] This sort of example is highlighted in Latus 2000.

[4] I suspect that an initial attraction to some sort of significance condition comes from focusing on the term "lucky," which, as I have already noted, comes with some fairly heavy affective connotations that might make it seem inappropriate to apply it to a mundane event. But putting the event in terms of its being a "matter of luck" serves to remove some of that pressure, I think.

control is a matter of luck for us. Period. This approach certainly has the virtue of theoretical simplicity. On the other hand, though, it complicates the evaluation of the theory by introducing an extra element of doubt into our intuitive assessments of cases. When we find ourselves reacting, say, with skepticism, to a putative case of luck, we would have to ask whether our intuition is sensitive to the proper application of the term or to the propriety of making a statement to that effect. This is notoriously difficult, as there seems always to be some Gricean appeal that preserves whatever judgment we prefer, and general principles of application are hard to come by.

Nevertheless, one or the other of these approaches (or some others I haven't considered) might well provide an adequate solution to the problem posed thus far for the control theory of luck, which is that insignificant events outside our control are not normally considered to be lucky, or even "matters of luck." But both strategies assume that *significant* events *uncontrolled by S* will always be a matter of luck for S. Unfortunately for the control theory, there are "sunrise" cases in which a radically uncontrolled (by S) yet very significant (for S) event appears not to be the least bit lucky (for S). I will develop just such a case, but I will do so in two stages. Consider first the following case.

Accidental Slayer

Emilia has been running away from a gang of vampires throughout the night. They have chased her all through her home town. They have been so close on her heels that she has been unable to stop to ask anyone for help. She eventually reaches the edge of town and begins to run across an empty field. Finally, though, she reaches her limit and stumbles exhausted to the ground. Inevitably, the vampires catch up to Emilia and immediately go in for the kill. Just at that moment, though, the sun peeks over the horizon, catching all the vampires in the open. They all burst into flame (as vampires must do at the touch of the sun's rays), and Emilia is saved. She gasps in unexpected relief.

This is a variant of a "sunrise" case in which the rising of the sun turns out to be vitally significant to Emilia (and mortally so to the vampires). It would appear that Emilia is lucky—lucky to be alive and lucky that the sun came up when it did. What this case shows is that sometimes mundane "background" sorts of events like the sunrise can be (or at least *could* be) significant. Hence, they can qualify as lucky on a control theory of luck with a significance condition. So this case does not pose any particular problem for the control theory of luck. But now consider a further variant.

Savvy Slayer

Francesca has been running away from a gang of vampires all night. They have chased her all through her home town. Francesca is very knowledgeable about these particular vampires, though. She knows their limits—how fast they can run and how long they can keep it up. They have remained right on her heels because she has made sure that she stays just out of their reach all night long, keeping them caught up in the frenzy of the chase. She times her sprint into the field to coincide with the dawn. She turns and stops to confront them just at the moment the sun peeks over the horizon, catching all the vampires in the open. They all burst into flame (as vampires must do at the touch of the sun's rays), and Francesca chuckles in satisfaction.

Here again we have a case in which the sunrise is, as usual, uncontrollable and also highly significant for Francesca (and the vampires). Yet in this case, I think, nobody would be tempted to say that Francesca was lucky— either to be alive or that the sun came up. (Or, to put it less tendentiously, that the rising of the sun was a matter of luck for Francesca.) Quite the contrary, in fact! (I, for one, would not dare to suggest to Francesca that the success of her plan was a matter of luck for her.) But this time, the control theory of luck is in trouble, since the sun's coming up meets both the LCR and the significance condition. How can the control theorist account for Francesca's lack of luck?

The way forward here for the control theory of luck is difficult. We can put the problem in the form of a dilemma: the luck theorist must either acknowledge that some uncontrollable events are nevertheless not a matter of luck or else find a way to plausibly expand the notion of control to somehow allow that Francesca really did have the requisite sort of control over the rising of the sun. Neither of these options looks enticing.

On reflection, though, there is something that does seem to be under Francesca's but not Emilia's control—the *coincidence* of the rising of the sun and their respective presences in the open field outside town. This conjoined state of affairs is what Francesca intentionally engineers with her night's work, while Emilia simply (and literally) stumbles into it. Here is the difference that seems to make a difference regarding the presence or absence of luck in these two cases. This coincidence is a matter of luck for Emilia but not for Francesca. If the control theory can give a plausible account of this difference, it might be salvageable from the threat of this kind of sunrise case.

Of course, even if it could, we would still have all the problems listed in section 2. For instance, even if we had a satisfying account of why the coincidence of the sun rising and the vampires' location was a matter of luck for Emilia but not for Francesca, we would still have to figure out whether the rising of the sun, *simpliciter*, is a matter of luck for either or both of them. The control theory with a significance requirement would still have it that it was a matter of luck for both Francesca and Emilia that the sun rose when it did. It is not obvious that the answer should be the same for the two of them. And is either of them lucky to be alive? Surely Emilia is. But how could the control theory account for this? The coincidence may be clearly a matter of luck, but it is not so clear how luck gets "spread around" the various components and aspects of the state of affairs, its causal history and its causal consequences.

This is clearly a problem for the control theory as a theory of luck generally. For even if the control theory has a good account of what makes coincidences a matter of luck, sunrise cases still seem to cause problems for it as an account of the many instances of luck (or its absence) that are not obviously characterizable as coincidences. Lucky coincidences are all instances of luck, but not all instances of luck are coincidences.

This need not necessarily be a problem for our theory of knowledge, however. It is worth noting that the luck generally involved in analyses of knowledge takes the form of a coincidence. One could capture the luck intuition about knowledge like this: "When it is a lucky coincidence that S believes that p and p, S does not know that p." This holds out the possibility that a less than fully general account of luck might be sufficient to inform a theory of knowledge in the way that luck theorists would like. To repeat, developing a satisfying general theory of luck would still be an eminently worthy project, but not one that must necessarily precede a satisfying theory of knowledge in terms of luck.

It will not help, however, to talk in terms of "lucky coincidences." The idea is to focus on a subclass of lucky phenomena so as to avoid requiring a fully general theory of luck. Obviously, then, we can't appeal to the general notion of luck itself when characterizing the relevant phenomena. We need a way to specify the subclass of luck phenomena without using the term "luck." I propose that we begin with the idea of a "mere coincidence." This does not appeal to a general notion of luck, at least at first blush.

4. Mere Coincidence

The idea of a mere coincidence is a familiar one, if not quite so familiar as that of someone's being lucky. To say that the coincidence of A and B is "mere" is to say that A and B are completely independent of one another—that one had "nothing to do with" the other. If A and B "merely" coincide, then there is nothing further relevant to say about their coincidence other than it occurred.

One way to get more clear about what we mean by a mere coincidence is to ask what kinds of connections between A and B defeat the "mere-ness" of the coincidence. The most intuitively obvious such connection is a causal one. If, for example, A caused B, then their coincidence was not mere. It's no (mere) coincidence that there is rising unemployment and falling consumer demand. It's no (mere) coincidence that the co-worker who was grotesquely obsequious to the boss got the promotion.

Nor is it a mere coincidence when A and B have a common cause. It is no (mere) coincidence that ocean levels are rising and severe weather events are increasing in frequency. And a particularly strong defeater of mere-ness is common cause by design. Any coincidence that is engineered by an agent is particularly non-mere. It is no (mere) coincidence that both Aisha and Blaise were at the party—Claudio invited them both.

I have been speaking as though the mere-ness of coincidence is a matter of degree rather than all or nothing. I think this is intuitively correct. Consider that in a fairly deterministic universe everything has a common cause if you go back far enough. Yet we are willing to say of a great many coincidences that they are mere coincidences. Imagine two mountain peaks on opposite sides of the earth, whose ranges share no geological history for millions of years. Now suppose that the two peaks are precisely the same height above sea level. This seems like a strong candidate for a mere coincidence. Yet if we go back far enough in time, there will be a set of causes that led to the current heights of the two peaks.

Contrast this with the case of a copse of trees, all of whose trunks are bare of branches up to precisely the same height. This is due to the presence of foragers that live in the forest and eat the foliage up to the height of the tallest forager. In this case, it is no coincidence that the trunks are bare up to the same point. But why not? Presumably because of the relative immediacy of the common cause. But this is, of course, merely a matter of degree.

Now consider a flower garden on a college campus full of hundreds of flowers of precisely the same species, size, and color and spaced precisely the same distance apart from one another. These facts are not remotely coincidental, because it is the result of painstaking design, intention, and effort together with the causal efficacy to bring about those intentions and designs.

And finally consider the possibility that an all-knowing and all-powerful deity has designed and brought forth the universe in all of its exquisite detail. In such a case, it would seem that nothing at all is merely coincidental. And, indeed, people who hold such views about the provenance of the universe say such things. These considerations suggest a rough spectrum of degrees of the mere-ness of coincidence, ranging from far-distant causal connections at one end to causally efficacious design and agency at the other. This gives us some purchase on the intuitive idea of mere coincidence. I do not have a developed theory of knowledge in terms of mere coincidence, but there are some good reasons to think that it is worth pursuing such a theory.

To begin with, most of the motivations for pursuing a theory of knowledge in terms of a general theory of luck also apply to doing so in terms of mere coincidence. Recall that a "lucky guess" is a paradigm case of a belief that does not count as knowledge. But S's lucky guess that p is simply the mere coincidence of S's believing that p and p being true. It's the same phenomenon being described, but the latter description is more specific about the way in which the guess is lucky. It's not that S is lucky to have entertained the thought that p, or that the world was such that p. S was lucky that the two happened together, which is to say that there is a kind of dependence (that is, lack of independence) between the two that obtains when S actually knows that p that is absent when he does not.

Gettier cases also can be re-described in terms of mere coincidences. In Gettier cases, it is a mere coincidence that (a) S has a justified belief that p and (b) that p is true. This is slightly different from a lucky guess, because by hypothesis lucky guesses are not justified. The kind of coincidence involved is thus slightly different. Nevertheless, Gettier cases are cases in which S justifiedly believes that p, yet despite this it is a mere coincidence that p is true. Again, there is an independence between the two states of affairs that both constitutes the mere-ness of the coincidence and undermines knowledge.

The idea of mere coincidence also has conceptual connections to the notions of achievement and credit similar to those that the idea of luck has. If the co-occurrence of A and B is a mere coincidence, then it does not count as anyone's achievement, and no one deserves credit for it.[5] And if we have robust enough intuitions about mere coincidences that are independent of our intuitions about knowledge, then we also preserve the benefits of a more top-down approach to epistemological theorizing, with all the attendant benefits to this alluded to in section 2.

And a theory of (mere) coincidence looks to be able to avoid some of the major difficulties that plague more general theories of luck. For a start, to say that some conjunctive state of affairs (A&B) is a mere coincidence is to say neither that (A&B) is positive (or negative) or that it is particularly significant. So when intuitively assessing whether or not (A&B) counts as a mere coincidence, we need not be distracted by considering whether (A&B) makes much difference to either the world at large or to any particular individual. While the significance condition may well be a necessary part of any general theory of luck, it has never struck me as a central feature of the kind of luck implicated in the theory of knowledge. So moving to a coincidence-based theory from a luck-based theory will allow us to divorce ourselves and our intuitions from this distracting consideration.

There is also the more general point that the term "coincidence" or even "mere coincidence" is not nearly so rife with associations with related terms. It seems to be a much less ambitious term than "luck" or "lucky," applying to a much more circumscribed set of circumstances and hence less easily conflated with others in the family tree (chance, fortune, and the like).

---

[5] There are potentially complicating issues here about whether mere coincidence can or must be indexed to individuals. For instance, in a general theory of luck, one must talk about an event's being "lucky-for-X" because the same event can be lucky for one person but not for another. On the control theory, this is explained by the fact that an event can be subject to one person's control but not to another's. Does it similarly make sense to talk about the co-occurrence of A and B being a mere coincidence for X but not for Y? I think intuition speaks less forcefully here than when we put this in terms of luck. It may well be necessary to make such a distinction in order to get particular cases right.

Of course, the down side of this is that it is not always applicable where we want a suitable judgment. Consider the Russian Roulette case. Rescher takes this case to show that lucky events need not be improbable, whereas Coffman argues that the survivor is not actually lucky, merely fortunate. Can an appeal to coincidence help decide this matter? Not really. There are coincidences one could examine in the situation. There's the coincidence that, say, chamber 1 had the bullet and chamber 3 is the one that was fired. Is that a mere coincidence? I don't know, but it hardly matters since it is not at all clear how to get from an answer to that question to an answer to the original question, which is whether the survivor is lucky to have survived. But this limited application of the concept of mere coincidence is not troubling if all we want to do is use it to help explain the theory of knowledge, in which the relevant phenomenon will always be the coincidence of belief and truth.

5. Conclusion

I have argued that there are good reasons to think that pursuing a theory of knowledge by way of a theory of the relevant kinds of luck that can undermine knowledge is a fruitful strategy. The recent explosion of interest in this strategy has yielded a lot of insight into both knowledge and luck. It is well worth developing a general theory of luck, given its centrality not just to our concept of knowledge but also to all sorts of important notions like responsibility, achievement, credit, freedom, and so on. But sustained attention to such a theory of luck has shown that it is a huge and complex subject matter in its own right. I propose that, for the purposes of understanding the specific way in which luck interacts with belief and justification with respect to knowledge, we attempt to focus on just the features of luck that are required, so that we can make progress on that front without having to defend a general theory of luck on all *its* fronts at the same time.

References

Ballantyne, Nathan. 2014. "Does Luck Have a Place in Epistemology?" *Synthese* 191:1391–1407.

Coffman, E. J. 2007. "Thinking About Luck." *Synthese* 158, no. 3:385–98.

———. 2009. "Does Luck Exclude Control?" *Australasian Journal of Philosophy* 87:499–504.

Greco, John. 2010. *Achieving Knowledge*. Cambridge: Cambridge University Press.

Lackey, Jennifer. 2008. "What Luck Is Not." *Australasian Journal of Philosophy* 86:255–67.

Latus, Andrew. 2000. "Moral and Epistemic Luck." *Journal of Philosophical Research* 25:149–72.

Pritchard, Duncan. 2005. *Epistemic Luck.* Oxford: Oxford University Press.

Rescher, Nicholas. 1995. *Luck: The Brilliant Randomness of Everyday Life.* Pittsburgh: University of Pittsburgh Press.

Riggs, Wayne D. 2007. "Why Epistemologists Are So Down on Their Luck." *Synthese* 158, no. 3:329–44.

———. 2009a. "Luck, Knowledge, and Control." In *Epistemic Value*, edited by Adrian Haddock, Alan Millar, and Duncan Pritchard, 204–21. Oxford: Oxford University Press.

———. 2009b. "Two Problems of Easy Credit." *Synthese* 169:201–16.

Zagzebski, Linda. 1999. "What Is Knowledge?" In *The Blackwell Guide to Epistemology*, edited by John Greco and Ernest Sosa, 92–116. Oxford: Blackwell.

CHAPTER 11

THE UNBEARABLE UNCERTAINTY PARADOX

SABINE ROESER

1. Introduction

People can be risk seeking and risk averse, but they can also be uncertainty averse.[1] In this essay I argue that uncertainty aversion can give rise to a puzzling phenomenon, which I wish to call the Unbearable Uncertainty Paradox. This arises when people are so uncertainty averse that they prefer a certain worst case to an uncertain state. Consider the following examples.[2]

Angus: Angus has symptoms that indicate he might have cancer. He visits the hospital, where he is subjected to various tests. He has to wait several weeks for the test results. Of course, Angus deeply hopes that he does not have cancer. When the results come in, they are inconclusive. More tests are needed. He has to wait a month for the test results. Halfway through this long period of uncertainty, Angus cries out: "I'd rather know that I have cancer now, instead of having to wait any longer for the test results."

Beth: Beth has applied for a big research grant. Her career depends on this grant. If she gets it, she will have a secure position in academia, which is what she wants. If she does not get it, her academic career is over. The referee reports are delayed. When they are sent, they are the wrong ones. The decision is postponed again and again. Eventually, she exclaims, "I'd rather know that I did not get the grant and find an alternative career than spend another minute with this unbearable uncertainty."

I think that these examples are very common. They indicate that next to the well-known phenomena of people being risk seeking and risk averse,

---

[1] E.g., Epstein 1999; also cf. literature on the Ellsberg paradox and ambiguity aversion, cf. Ellsberg 1961.

[2] I owe these examples and their formulations to Martijn Blaauw.

The Philosophy of Luck, First Edition. Edited by Duncan Pritchard and Lee John Whittington.
Published 2015 by John Wiley & Sons Ltd.

people can also be uncertainty averse: in other words, if risk is at least the possibility of an unwanted effect, then it is not only the unwanted effect that they want to avoid, it can also be the uncertainty inherent in the possibility that they wish to avoid. But these examples also reveal something about our psychology that is interesting and puzzling. They indicate that uncertainty aversion can lead to a state where someone prefers a certain outcome at all costs, even when it is the worst case. Apparently, in situations of uncertainty that relate to deeply existential issues such as health and career, people can reach a point where uncertainty has such an inherent disvalue that this can overshadow the hope for a positive outcome to a degree that people prefer the worst case, and certainty. This, however, gives rise to the following paradox: the worst case seems to be more acceptable than the state where it has not yet materialized. This is what I call the Unbearable Uncertainty Paradox. This phenomenon seems to be widespread but nevertheless has not been identified before, so far as I know. This essay provides a first sketch of this phenomenon and how it might be analyzed from different approaches to risk and rationality.

2. A Formal Analysis of the Unbearable Uncertainty Paradox

In order to understand the paradoxical nature of the phenomenon of the Unbearable Uncertainty Paradox (UUP), I will start by presenting a formal analysis.[3] To that purpose, I analyze the UUP in terms of conflicting preference orderings.

I understand "to prefer" as a three-place relationship: "S prefers x over y." There is a paradox if "S prefers x over y" and "S prefers y over x" at the same moment (with y and x not being on a par). Let "$K(p)$" denote "to know that p." Note that $K(p)$ entails that p is true (with knowledge understood as at least justified *true* belief).

Normally, Angus's preference ordering (PO) is as follows:

PO1. $K(\text{not } p) > \text{not } (K(p) \text{ or } K(\text{not } p)) > K(p)$

Angus prefers to know that he does not have cancer (with "p" denoting "I have cancer") to knowing that he has cancer. He normally prefers the uncertain state where he does not know either way to the state where he does know that he has cancer, because in the uncertain case, he still can hope. However, when Angus falls prey to the UUP his preference ordering changes to:

PO2. $K(\text{not } p) > K(p) > \text{not } (K(p) \text{ or } K(\text{not } p))$

[3] I would like to thank Martijn Blaauw and Francine Dechesne for very helpful comments on earlier versions of my formalization.

This preference ordering expresses the idea that the uncertainty is so tormenting that Angus prefers to know that p is true, so that he can regain his agency rather than having to bear the uncertainty any longer.

The paradox, then, is that even during the UUP Angus still also holds the preference ordering PO1; he is "of two minds." An alternative explanation would be that he swings back and forth between the two preference orderings, PO1 and PO2. In that case, there would not be a logical paradox. It is not clear, however, that this is really the case when somebody experiences unbearable uncertainty. Furthermore, even if it were to happen, this would presumably be due to the fact that it would be a paradox if Angus would hold both preference orderings at the same time.

Is this trivial—that is, is falling prey to the UUP simply a sign of irrationality and hence not worth any serious discussion? Or is the UUP a widespread phenomenon among otherwise rational persons that deserves discussion in order to understand why they fall prey to it? If so, is it another example of bounded rationality and heuristics and biases (forms of irrationality that are pervasive among normal people) or might there even be something rational, or at least understandable, behind this apparent paradox? These are the questions I address in the remainder of this essay.

3. Heuristics, Biases, and Beyond

Over the past few decades, a lot of psychological research has been done on risk perception and decision making under uncertainty. The most influential school of thought is the one founded by Amos Tversky and Nobel laureate Daniel Kahneman (Tversky and Kahneman 1974; Gilovich, Griffin, and Kahneman 2002; also see Kahneman 2011 for a popularized account). This research provides for empirical evidence that people are very bad at processing statistical information and at logical reasoning. Scholars working in this area argue that we have two distinct systems with which we can process information, a theory commonly called Dual Process Theory. People are prone to rely on intuitive processes of thinking ("system 1") that provide them with fast heuristics which help them navigate smoothly through a complex world but which are very unreliable. Analytical, rational thinking ("system 2") is more reliable but requires more time and resources. System 1 is supposed to be emotional, intuitive, unconscious, and irrational, whereas system 2 is supposed to be analytical, deliberative, conscious, and rational. This work is extremely influential in psychology and empirical decision theory, and recently more and more philosophers have also become interested in this approach. Based on this account one could argue that the UUP is just another example of our limited capacities to appropriate decision making under uncertainty. On this interpretation, the UUP arises through our supposedly intuitive and subconscious system 1, but analytical system 2 evaluation reveals that this is paradoxical and irrational.

Our philosophical story could end here; we could leave it up to psychologists to gather empirical evidence of the prevalence of this phenomenon. However, there are scholars in psychology and philosophy who are critical toward Dual Process Theory. The division of our information-processing capacities into two distinct systems gives rise to various philosophical and conceptual problems (Roeser 2009, 2010). Not all biases are emotional, not all emotions are biases, and not all supposed biases really are biases (Roeser 2010). Not all intuitive processes are irrational and emotional, such as insight into mathematical axioms, and not all emotions are unconscious and spontaneous, such as the love for our family and friends and various moral emotions that have cognitive content, have a narrative structure, and allow for reflection and deliberation (Roeser 2009). Ethical intuitionists have developed accounts that show that even unreflective and spontaneous moral intuitions can be justified and justifiable (e.g., Reid 1969 [1788]; Ross 1968 [1939]; Broad 1951 [1930]).

In addition, there is empirical evidence that intuitive processes can actually outperform and be more reliable than analytical processes, especially when it comes to expert knowledge (Gigerenzer 2007). This work can be related to the distinction in philosophy between "knowing how" and "knowing that." After having learned to ride a bike we automatically know how to do so, but if asked we might find it difficult to articulate what we are doing (propositional knowledge that).

This gives rise to the following question: Is the UUP really a form of irrationality, or might the UUP be another instance of apparent irrationality that turns out not to be so irrational in all cases? Or is the UUP at least an understandable phenomenon? Does the UUP provide any guidance and help, is it a heuristic? In order to answer this question I first, in section 5, provide for a taxonomy of the UUP. What are typical examples, and in what kinds of situations does it occur? This is followed by an analysis of the phenomenology of the UUP in section 6. In section 7 I then discuss how rational or irrational the UUP is.

4. A Taxonomy of the UUP

I will discuss three major areas in which the UUP can occur, concerning the following types of risk: type 1, health risks; type 2, voluntarily invoked life-goal risks; and type 3, technological risks.

I've already given examples for types 1 and 2: Angus's case is an example of the UUP concerning a health risk (type 1), and Beth's case is an example of the UUP concerning a voluntarily invoked risk, related to one's life goals (type 2). An example of type 3, a technological risk, would be the risk of a nuclear meltdown, for example the imminent danger in the aftermath of the Fukushima tsunami. The UUP is less likely to manifest itself in an everyday situation; in all three examples, there is a direct existential threat. People can always worry about their health, risky technologies, and their life goals. In those cases, they will have their normal preference orderings, similar to Angus's PO1 However, in the case of imminent revelation of information concerning a preferred outcome or a worst case, people may experience the tension of that revelation as unbearable and slide into PO2, while still endorsing PO1, and hence undergo the UUP. So what the three types have in common is that a lot is at stake, and that the outcome is imminent, but it might also be uncertain when exactly the outcome will manifest itself, so there can be uncertainty concerning both the outcome and the timing of the outcome, which can make the uncertainty even more unbearable.

Let us now see what distinguishes these cases, which can provide us with a taxonomy of UUP cases and a further understanding of the UUP. To that purpose, I will look at several of the salient qualitative dimensions of risk that are commonly distinguished in the literature on ethical and social aspects of risks (Slovic 2000; Roeser 2006, 2007; Asveld and Roeser 2009). These dimensions are whether risks are collective or individual, how risks and benefits are to be assessed, measured, compared, and distributed, what the status quo is, voluntariness, and natural risks versus human-made risks. These dimensions are not part of formal approaches to risk in rational decision theory, but they play a major role in people's risk perceptions, as we can learn from empirical decision theory (Slovic 2000). From an ethical point of view, these dimensions can be normatively relevant (Roeser 2006, 2007; Asveld and Roeser 2009; Shrader-Frechette 1991; Hansson 2004). For example, cost-benefit analysis, the dominant approach in formal methodologies for risk assessment, focuses on collective risks and benefits and does not pay attention to the impact of risks and benefits for individuals. By focusing on aggregate levels of risks and benefits, it also overlooks issues of equity and fair distribution of these risks and benefits within a given population. It is often far from clear how and on what scale risks and benefits should be measured and compared (cf. Espinoza 2009 on the problem of incommensurability in the context of risk). Attachment to the status quo can be a bias when the status quo is bad and can easily be improved at no risk, but if there are good reasons to value the status quo and a new development can provide for improvements but also for new risks, there might be good reasons to be cautious about these developments. Voluntariness relates to the morally important concept of autonomy, but it does not figure in the dominant consequentialist, technocratic approaches to risk, such as cost-benefit analysis. Naturalness tends to be an important factor in laypeople's risk perception, but it is not distinguished as such in formal risk approaches. This aspect might be more contentious. Naturalness can be more coherent with a sustainable, health-conscious lifestyle. On the other hand, that something is natural does not necessarily mean that it is less risky than something that is artificial or technology based (cf. Hansson 2003). There are many dangerous substances to be found in nature, such as uranium and asbestos, and there are artificial substances that are not dangerous; furthermore, technologies can protect us from natural hazards, such as dikes and walls. In the following discussion I analyze to what extent these dimensions play a role in the three different types of risk.

Collective risks versus individual risks. Health risks are predominantly experienced individually, with the exception of pandemics. Life-goal risks are typically individual as well. In contrast, technological risks such as risks related to energy production, infrastructures, climate change, or the use of new substances such as nanoparticles are often collective, due to the large scale on which they are introduced into society. In the case of collective risks, the options for action for individuals are less clear than in the case of voluntarily invoked, individual life-goal risks. This relates to the dimension of voluntariness that I discuss below.

Risks versus benefits. In risk types 1 and 2 the benefits of a possibly risky situation are clear. Good health is a condition that everybody aspires to; possible illness is an unfortunate reality that many people have to deal with. Getting a major research grant is a very desirable situation, hence it might be worthwhile undergoing the risk of not getting it. However, in type 3 the benefits are often less clear to society, which explains why the risks are seen as more controversial. Types 1 and 3 have clear negative impacts; in the case of risk type 2, the impact can be as in Beth's case, where a negative outcome means that one has to find a different career, but it can also be less dramatic, where a new job or grant can improve one's current work situation although it is acceptable as it is.

Status quo. Another important question when thinking about the acceptability of risks is what the status quo is. In type 1 the status quo is the best case; the status quo is that people are healthy. Medical treatments serve to restore the status quo. In type 3 the status quo is to do nothing; the perception in cases of controversial technologies is often that they pose seemingly unnecessary risks. Risky technologies are developed to improve well-being, but that is usually an improvement of the status quo. The widespread phenomenon of status quo bias (cf. Samuelson and Zeckhauser 1988) explains the aversion people have against such technologies: the status quo is seen as the normal or natural state. Sociological studies of technology acceptance show that currently uncontroversial technologies have often been perceived as threatening and dangerous. Similar patterns occur again and again when a new technology is introduced (cf. Schivelbusch 1986 [1977]). It can be a misconception, however, that the status quo is risk free (cf. Sunstein 2005). On the other hand, there are technologies that turn out to be very risky. Once they are introduced into society, they develop a momentum of their own, such as nuclear weapons, which played a crucial role in the arms race during the Cold War, and nuclear energy, which has created a still unsolved problem of nuclear waste (cf. Taebi, Roeser, and Van de Poel 2012). In the case of type 2 the status quo is the worst case, so we have nothing to lose. In sum, then, in type 1 the status quo is the best case, in type 3 the status quo tends to be perceived as best, even though this might be unjustified, and in type 2 the status quo is the worst case.

Controllability, voluntariness, and natural risks versus human-made risks. I will discuss these notions together as they are closely related in the context of my discussion. All three types of risk have in common that they are characterized by an experience and factual lack of control. However, while possible cancer patients probably recognize a UUP expressed in Angus's story, in the case of technological risk, it is not as clear that a person would say: "I'd rather have the certainty of a nuclear meltdown than the uncertain risk of a meltdown." This asymmetry might be explained by the following difference between the two cases, namely, the even more involuntary nature of health risks versus the human-made, and in that sense avoidable, risks of nuclear energy. In the latter case, people might rather wonder why we put ourselves in this situation in the first place. Because health risks are to a certain degree natural and unavoidable, there is a need for coping and fatalism. In the case of technological risks, fatalism is less appropriate because as a matter of fact they are avoidable. On the other hand, to the extent that technological risks result from collective actions, individuals might still feel helpless and that they lack control. In the case of collective action problems, we have to deal with the complex interrelationships between individual and collective stakes, benefits, and responsibilities. Hence, also in type 3 cases people might experience the UUP in a way similar to the way they do in type 1. Also, in the case of an imminent threat, issues of causal responsibility might be less relevant, as they do not make a difference at that stage. This becomes clear in the case of type 2, where we experience the UUP even though the uncertainty may have been partially initiated by ourselves, but at this moment it is out of our hands, as in the case of Beth, who has to wait for the decision of others concerning her fate.

Let us complicate matters a bit by acknowledging that distinguishing the cases according to degrees of voluntariness is problematic. For example, an unhealthy lifestyle can cause health problems. Lifestyle-induced health risks can then be seen as a mix of types 1, 2, and 3: they concern health (type 1), are self-invoked (type 2), and are human-made and can involve technology (type 3). The link between lifestyle and health risks, however, is less evident in specific cases than in cases of type 2, as health problems can always be due to multiple causes and it can be hard to trace back in individual cases what the contribution of lifestyle was to a health problem. Also, cases of type 2 are only partially initiated by ourselves, as arguably the circumstances can force us to engage in uncertain activities. This holds in the case of Beth, who has to try to get the grant in order to obtain a permanent position in academia. On the other hand, one could argue that it was her choice to try to get into academia in the first place. This illustrates that individual choices and systemic circumstances are intertwined and it is a bit artificial to disentangle them. I do so nonetheless, however, as the

TABLE 1. A taxonomy of UUP types and their qualitative dimensions of risk

	Type 1 (health)	Type 2 (life goals)	Type 3 (technology)
Collective	−	−	+
Value of benefits	+	+	−
Status quo attachment	+	−	0
Voluntariness	−	+	0
Human-induced risk	−	+	+
Controllability	−	−	−

different cases are more or less voluntary to different degrees, albeit not purely one way or the other.

Table 1 summarizes this taxonomy. We can see that, on the one hand, the UUP can occur in all three types of risk, and a feature they all share is a lack of control. On the other hand, the UUP is presumably more likely to occur in types 1 and 2. In both types, the UUP is motivated by an urge to settle with fate. The difference is that in type 1 the situation is out of one's hands from the start, whereas in type 2 the situation was initially in one's own hands, when trying to achieve an aspired life goal, but now is out of one's hands, while one has to wait for the external decision. In the case of type 3, the UUP is less common. More likely responses are presumably resistance to the technology in the first place, possibly in the form of a "not in my backyard" response. However, the UUP can occur in the face of imminent danger, for example when there is a direct threat of a nuclear meltdown. The differences in the qualitative dimensions of these types of risk make them more prone or less prone to the UUP.

5. The Phenomenology of the UUP

What seems to characterize the UUP is the experience of an unbearable psychological pain. This is a pain that at that moment is perceived to be worse than the projected experience of the worst case. At the same time, this is accompanied by an awareness of the irrationality of this state or preference ordering (PO2), as we are also aware of our normal preference ordering, PO1 We could characterize this as the difference between dispositional versus occurrent preferences, analogous to the differences between dispositional and occurrent beliefs (cf., e.g., Schwitzgebel 2010) and between dispositional and occurrent emotions (Wollheim 1999). Another helpful distinction taken from epistemology concerns seemings versus considered judgments. In this respect, it can be helpful to compare the UUP to the case of perceptual illusions.

The *analogy with perceptual illusions* is as follows. When we are prone to a perceptual illusion, we see something in a certain way x, while knowing that x is not the case. In philosophy, this phenomenon is discussed in the literature on so-called seemings. Seemings are seen as nondoxastic states that are distinguished from considered judgments with which they can conflict while nevertheless persisting (cf. Szabó Gendler 2008; Cullison 2010). In psychology, perceptual illusions are discussed yet again in the context of heuristics and biases; they are seen as a clash between system 1, namely, an illusion based on an unconscious gut reaction, and system 2, a considered judgment or rational belief (cf. Sloman 2002). In the case of the UUP, we can understand this as a clash between an occurent emotion or first-order desire, namely, PO2, based on the painfulness of the uncertainty, and a dispositional emotion or second-order desire, namely, PO1, based on our considered judgment.

At the same time, however, there is a *disanalogy with perceptual illusions.* In normal circumstances, perceptual illusions can fairly be characterized as an intellectual puzzle that can be resolved by rational explanation. Even though the seeming might persist, we know that it is mistaken, and this more or less settles the matter. The UUP is not, however, an intellectual puzzle but an existential struggle. Rather than being resolved by rational reflection, this struggle can even be reinforced by reflection, because we get more and more absorbed by the object of our anxiety. This means that the struggle gets worse, and we cannot let go of it; it can even become an obsessive anxiety. The well-meant advice to people who suffer from situations in which they experience the UUP—"Don't think about *x*"—might not help at all or might even lead to more thinking about *x*. In the case of the UUP, the best we might be able to achieve is to be able to cope with the uncertainty (cf. Little and Halpern 2009 on coping).

What we deeply wish in such a situation is to regain control. *Withdrawal* from the situation can then be experienced as a solution. But in the cases I have discussed, withdrawal is not a real option or not an attractive option.

For example, if Beth would withdraw her funding application, she would never know whether she might not have gotten the grant after all, and this would mean that she would act in accordance with her occurrent preference ordering, PO2, but thereby violating her dispositional, higher-order, considered preference ordering, PO1. She might, however, learn via the UUP that PO1 comes at too high a price—namely, the uncertainty involved that is too hard for her to bear. In that case, she might revise her preference ordering, settle for PO2, and indeed withdraw from the funding competition, thereby regaining control over the situation. There are cases, however, in which people suffer so severely from UUP stress that they cannot stand the pressure anymore; the conundrum between PO1 and PO2 is presumably unsolvable. Indeed, there are cases where people commit suicide while waiting for the result of a competition. In France, Michelin-star cook Bernard

Loiseau committed suicide shortly before he was to hear whether he would keep his three Michelin stars, because of negative reviews and rumors that he might lose one of his stars. The pressure became unbearable for him. In the end, his restaurant retained the star.[4]

In the case of Angus, withdrawal is not even an option, unless it is in form of suicide. A less radical, yet still extreme response would be for him to withdraw from medical diagnosis altogether, perhaps turning to alternative medicine.

In the case of technological risks, there are people who try to withdraw and lead an alternative lifestyle. One of the intricate ethical aspects of many technological risks, however, is that they can also affect people who choose not to make use of the risky technology. Technological risks can have far-reaching effects, for example on the environment or through accidents that can also affect people who are not directly involved in using the technology.

Hence, in some cases it might be possible to withdraw from the UUP in order to regain control; in other cases, coping might be the best available option. In the case of technological risks, joining an environmental movement or trying to change policy or technology development in other ways might be other alternatives.[5]

6. How Irrational Is the UUP?

The question remains whether the UUP is an irrational, biased state that needs to be corrected, or whether it is an understandable phenomenon that might even reveal insights that we might not be aware of without it. I think that the UUP is ambiguous in that respect, as I argue in what follows.

Time, perspective, and context are crucial factors for understanding the UUP. Retrospectively or from an outsider perspective, the UUP is a blatant paradox. For that reason the UUP can appear as manipulative or dishonest. When one is inside the UUP, however, the existential tension is inescapable; one can get into a loop and be neurotically obsessed with the feared outcome. From a detached, objective point of view the UUP is paradoxical and irrational, but from a narrative, understanding approach it makes sense despite its paradoxical nature. Intuitively, what makes it difficult to cope with uncertainty is that it is unclear which options the agent has for action, as the outcomes can go either way.

---

[4] Cf. http://en.wikipedia.org/wiki/Bernard_Loiseau for information. Thanks to Peter Kroes for drawing my attention to this case as an example of the UUP.

[5] For example, over the past few decades many approaches for participatory technology assessment have been developed that try to include laypeople in constructive decision making about risky technologies and in technology development itself (cf. Van Asselt and Rijkens-Klomp 2002). Such approaches can help to overcome the UUP, specifically when it comes to long-term, nonurgent risks, but arguably involving the public is also a way to alleviate the burden of the UUP in the direct onset of a disaster.

From an evolutionary perspective, we are not prepared for any of the three types of risk that I have discussed:

Type 1, health risks: in the past, maybe even a hundred years ago, we had no (sound) knowledge about health risks; sophisticated medical diagnoses as we know them today did not exist to any significant extent.

Type 2, voluntarily invoked life-goal risks: these risks are something to which humans have always exposed themselves, but in the past, this was a privilege for a small class of people, whereas in contemporary, individualist Western society, designing one's life and destiny is pervasive in all classes, albeit in different forms and to different degrees.

Type 3, technological risks: we have experienced such risks since the industrial revolution, but genuine technoscience, with concomitant risks and uncertainty, has also mainly developed over the past hundred years.

Hence, an evolutionary perspective can indicate that it is not surprising that we have difficulty dealing with risk and uncertainty, which can lead to conflicting preferences, stress, and anxiety.

In some cases, however, our difficulty in dealing with uncertainty can be a useful mechanism: it can open the path to coping. The manifestation or expectation of the worst case, bad as it may be, allows for settling, accepting, and coping. It gives the agent back her agency. By anticipating a bad outcome one can prepare an alternative strategy. The UUP becomes dangerous when it becomes a self-fulfilling prophecy, when one does not try to improve the status quo even if one can.

7. Conclusion

This essay has been rather speculative, providing a first sketch of a phenomenon that seems to be widespread but nevertheless, to the best of my knowledge, has not been identified before, neither by philosophers nor by psychologists. I hope the essay can make a humble contribution toward starting academic research on this topic. Empirical research would be necessary to study what the prevalence of the UUP is and to analyze its phenomenology in more detail. For now I think we can say that the UUP is a paradox and is in some sense irrational and even experienced as such, but nevertheless it is very common and understandable from a first-person perspective. Furthermore, the UUP can provide insights into coping, alternative strategies, and regaining one's agency in situations where one is exposed to existential threats.

Acknowledgments

The initial idea for this essay arose from conversations that I had with Martijn Blaauw. I would like to thank audiences at a philosophy colloquium at TU Delft in October 2012 and at a workshop on luck at the University of

Edinburgh in June 2013 for their helpful feedback on drafts of the essay. My work on the essay was funded by a VIDI grant for research on "Moral Emotions and Risk Politics" from the Netherlands Organization for Scientific Research (NWO; grant no. 276-20-012).

References

Asveld, Lotte, and Sabine Roeser (eds.). 2009. *The Ethics of Technological Risk*. London: Earthscan/Routledge.

Broad, C. D. 1951 [1930]. *Five Types of Ethical Theory*. London: Routledge and Kegan Paul.

Cullison, Andrew. 2010. "What Are Seemings?" *Ratio* 23:260–74.

Ellsberg, Daniel. 1961. "Risk, Ambiguity, and the Savage Axioms." *Quarterly Journal of Economics* 75, no. 4:643–69.

Epstein, Larry G. 1999. "A Definition of Uncertainty Aversion." *Review of Economic Studies* 66, no. 3:579–608.

Espinoza, Nicolas. 2009. "Incommensurability: The Failure to Compare Risks." In *The Ethics of Technological Risk*, edited by Lotte Asveld and Sabine Roeser, 128–43. London: Earthscan/Routledge.

Gigerenzer, Gerd. 2007. *Gut Feelings: The Intelligence of the Unconscious*. London: Viking.

Gilovich, Thomas, Dale Griffin, and Daniel Kahneman (eds.). 2002. *Heuristics and Biases: The Psychology of Intuitive Judgement*. Cambridge: Cambridge University Press.

Hansson, Sven Ove. 2003. "Are Natural Risks Less Dangerous Than Technological Risks?" *Philosophia Naturalis* 40:43–54.

———. 2004. "Philosophical Perspectives on Risk." *Techné* 8:10–35.

Kahneman, Daniel. 2011. *Thinking Fast and Slow*. New York: Farrar, Straus and Giroux.

Little, Margaret, and Jodi Halpern. 2009. "Motivating Health: Empathy and the Normative Activity of Coping." In *Naturalized Bioethics*, edited by Hilde Lindemann, Margaret Walker, and Marian Verkerk, 141–62. Cambridge: Cambridge University Press.

Reid, Thomas. 1969 [1788]. *Essays on the Active Powers of the Human Mind*. Introduction by Baruch Brody. Cambridge, Mass.: MIT Press

Roeser, Sabine. 2006. "The Role of Emotions in Judging the Moral Acceptability of Risks." *Safety Science* 44:689–700.

———. 2007. "Ethical Intuitions About Risks." *Safety Science Monitor* 11:1–30.

———. 2009. "The Relation Between Cognition and Affect in Moral Judgments About Risk." In *The Ethics of Technological Risk*, edited by Lotte Asveld and Sabine Roeser, 182–201. London: Earthscan/Routledge.

———. 2010. "Emotional Reflection About Risks." In *Emotions and Risky Technologies*, edited by Sabine Roeser, 231–44. Dordrecht: Springer.

Ross, W. D. 1968 [1939]. *Foundations of Ethics*. The Gifford Lectures. Oxford: Clarendon Press.

Samuelson, William, and Richard Zeckhauser. 1988. "Status Quo Bias in Decision Making." *Journal of Risk and Uncertainty* 1, no. 1:7–59.

Schivelbusch, Wolfgang. 1986 [1977]. *The Railway Journey: The Industrialization of Time and Space in the 19th Century.* Berkeley: University of California Press.

Schwitzgebel, Eric. 2010. "Acting Contrary to Our Professed Beliefs, or The Gulf Between Occurrent Judgment and Dispositional Belief." *Pacific Philosophical Quarterly* 91:531–53.

Shrader-Frechette, Kristin. 1991. *Risk and Rationality.* Berkeley: University of California Press.

Sloman, S. A. 2002. "Two Systems of Reasoning." In *Intuitive Judgment: Heuristics and Biases*, edited by Thomas Gilovich, Dale Griffin, and Daniel Kahneman, 379–96. Cambridge: Cambridge University Press.

Slovic, Paul. 2000. *The Perception of Risk*. London: Earthscan.

Sunstein, Cass R. 2005. *Laws of Fear*. Cambridge: Cambridge University Press.

Szabó Gendler, Tamar. 2008. "Alief and Belief." *Journal of Philosophy* 105, no. 10:634–63.

Taebi, Behnam, Sabine Roeser, and Ibo van de Poel. 2012. "The Ethics of Nuclear Power: Social Experiments, Intergenerational Justice, and Emotions." *Energy Policy* 51:202–6.

Tversky, Amos, and Daniel Kahneman. 1974. "Judgment Under Uncertainty: Heuristics and Biases." *Science* 185:1124–31.

Van Asselt, Marjolein, and Nicole Rijkens-Klomp. 2002. "A Look in the Mirror: Reflection on Participation in Integrated Assessment from a Methodological Perspective." *Global Environmental Change* 12:167–84.

Wollheim, Richard. 1999. *On the Emotions.* New Haven: Yale University Press.

CHAPTER 12

GETTING MORAL LUCK RIGHT

LEE JOHN WHITTINGTON

1. Introduction

In this essay I provide a modal account of moral luck, with a focus on resultant moral luck. The account is built upon the previous work of Duncan Pritchard (2005 and 2006) and Julia Driver (2013), who have also attempted to provide modal accounts of moral luck. Their accounts of moral luck, however, both require some revision on the basis of two problems.

The first of these problems is a problem of inclusivity. The current accounts of modal moral luck can potentially include too many kinds of cases as being cases of moral luck. I argue that this is due to a lack of further world-fixing conditions that would precisely pick out what kinds of cases qualify as cases of moral luck. In the same way that Pritchard's (2005) epistemic luck fixes the way the belief was formed across nearby possible worlds in order to assess whether there has been any knowledge undermining luck, I suggest as a solution to this problem that, for moral luck, it is the action performance that needs to be fixed across nearby possible worlds if we are to assess the results of an action as being morally lucky (rather than lucky in another kind of way).

The second problem I consider is a problem with the significance or interests condition for luck picking out the wrong value for cases of moral luck. Specifically, if we understand the significance condition in terms of interests of the agent, then it is possible to construct cases of moral luck that should be cases of good moral luck but come out as bad moral luck, or vice versa. My solution to this is to relativise the significance condition to the normative domain for which the luck is being attributed. For moral luck, the kind of significance that needs to be fulfilled is a moral

The Philosophy of Luck, First Edition. Edited by Duncan Pritchard and Lee John Whittington.
Chapters © 2014 Metaphilosophy LLC and John Wiley & Sons Ltd, except for "Luck as Risk and the Lack of Control Account of Luck" © 2015 Metaphilosophy LLC and John Wiley & Sons Ltd. Book compilation © 2015 Metaphilosophy LLC and John Wiley & Sons Ltd.
Published 2015 by John Wiley & Sons Ltd.

significance or moral value. What kind of value this is will depend on one's ethical theory.

Finally, I consider kinds of moral luck other than resultant moral luck and what these kinds of moral luck should look like if we are motivated to use a modal understanding of luck for these types as well.

2. What Is Moral Luck?

A paradigmatic case of moral luck is a case of two equally reckless truck drivers, driving on two equally safe roads. The first driver hits a pedestrian. The second driver does not. The argument goes that the first driver is more morally blameworthy than the second driver, despite the fact that there being a pedestrian in the first driver's path is beyond the driver's control. That is, the first driver drove recklessly and killed someone, which is morally worse than the case of the second driver, who only drove recklessly. This kind of moral luck is called resultant moral luck—getting morally lucky or unlucky with regard to results of one's actions (in this case, the results of reckless driving). There are at least three other kinds of moral luck, circumstantial, constitutive, and causal luck (Nagel 1993), each of which holds that an agent is morally blameworthy or praiseworthy (or avoids moral blameworthiness or praiseworthiness) due to factors beyond her control. For simplicity and unless otherwise specified, the focus of this essay is on resultant moral luck.[1]

There may be reasons to be suspicious about the existence of moral luck (Williams 1993). Part of the problem of moral luck is that intuitively we do not think that factors significantly beyond an agent's control should play a role in the agent's moral blameworthiness or praiseworthiness. The reason for this thinking is that factors beyond the agent's control are also beyond the agent's responsibility, and what one is not responsible for one should not be blamed for. To use our truck driver example again, the first driver should not be held responsible for injuring or killing the pedestrian, as the fact there was a pedestrian in his path was beyond his control. Note that this does not mean he is not responsible for reckless driving. Yet this thinking very often does not square with our practices of blaming and praising, where we are more likely to blame the first truck driver more than the second. This is not just a dilemma between intuitions and practices. We might simply hold that our practices are mostly in error, but the problem of moral luck will occur for any account of moral blameworthiness or praiseworthiness that holds that the outcomes of an action play at least some role in evaluating the goodness or badness of an action (Driver 2013). That is, any account with a glimmer of externalism about the goodness or badness of an action will fall prey to some form of moral luck problem. One might

---

[1] For some discussion on circumstantial moral luck see Hanna (forthcoming).

retreat again to a purely internalist account of moral goodness or badness, yet, as Julia Driver remarks, "When it comes to the significance of outcomes people will frequently note that the agent's impact on the world is morally significant—and to deny that significance encourages a kind of moral solipsism" (2013, 6). If this is right, then moral luck is here to stay.

Yet recent work on luck itself has begun to have an impact on what we think of as moral luck. Luck itself has for a long time been left undefined. Both of the seminal papers by Nagel (1993) and Williams (1981) around which most of the contemporary debates on moral luck revolve suffer from never clearly defining what moral luck, let alone luck itself, actually is. Williams unashamedly remarks that he will "use the notion of 'luck' generously [and] undefinedly" (1981, 22). However, luck is now a better-defined notion. We now have at least two general accounts of luck available—the lack of control account of luck and the modal account of luck.

3. The Lack of Control and Modal Accounts of Luck

Much of the discussion of moral luck has explicitly or implicitly rested on a lack of control account of luck (LCAL). For example, Daniel Statman writes of moral luck: "Let us start by explaining what we usually mean by the term 'luck'. Good luck occurs when something good happens to an agent P, its occurrence being beyond P's control. Similarly, bad luck occurs when something bad happens to an agent P, its occurrence being beyond his control" (1991, 146). And here is Michael Zimmerman on the same subject: "Something which occurs as a matter of luck with respect to someone P is something which occurs beyond P's control" (1987, 376). In order to be precise, we need a more formal characterisation of the lack of control account. Here is Jennifer Lackey's formulation of LCAL: "An event E is lucky for an agent S if and only if E is beyond, or at least significantly beyond, S's control and E is significant for S" (Lackey 2008, 256).

LCAL captures some genuine cases of luck. Winning a fair lottery is counted as lucky for the winner because the winning of the lottery is both beyond her control and significant for her (assuming some details about the financial state of the winner). What about cases of moral luck? In our truck driver case, it is beyond the control of our reckless driver that there is a pedestrian in his way, and it is significant for him that he hit the pedestrian.

Unfortunately, due to the more recent analysis of luck itself, LCAL has been shown to be a flawed general account of luck, with several different kinds of objections coming from different authors. The foremost problem, first raised by Andrew Latus (2000), is the sunrise problem. The sun rises each morning; this is beyond my control and this is significant for me, yet the sun rising each morning is not lucky for me. If this objection is not seen as fatal, other problems (without going into too much detail) include Lackey's (2008) Demolition Worker problem and Pritchard's (2005) epistemic problem—where LCAL will count too many true beliefs as being lucky—as well as problems associated with understanding what it means to be *significantly* beyond the control of an agent and whether it is only the lucky agent beyond whose control the event must be, or whether this extends to other agents as well. Even if all these problems were to be resolved, LCAL still misses something about what is generally regarded to be the case about lucky occurrences—that they are in some way chancy, improbable, unforeseeable, and/or unlikely to happen.

In an attempt to solve the various problems concerning epistemic luck, Pritchard (2005) develops an account of luck that captures the notion of chanciness as well as avoids the pitfalls encountered by LCAL. This account of luck is the modal account of luck (MAL), and it goes as follows:

> 1. If an event is lucky, then it is an event that occurs in the actual world but which does not occur in a wide class of the nearest possible worlds where the relevant initial conditions for that event are the same as in the actual world.

> 2. If an event is lucky, then it is an event that is significant to the agent concerned (or would be significant, were the agent to be availed of the relevant facts). (Pritchard 2005, 128–32)[2]

By appealing to regular occurrence in nearby possible worlds, the modal account captures the notion of chanciness. An event that could easily have been otherwise will be a lucky event, whereas "modally robust" events or events that could not so easily have gone otherwise will not be counted as lucky. Furthermore, MAL also mostly captures lack of control.[3] For the most part, an event under the control of an agent cannot also be a modally non-robust event.

We can see how this account avoids some of the problems that face LCAL. For example, consider the sunrise problem: the rising of the sun each morning is a modally robust event, as it occurs in the actual world and a wide set of nearby possible worlds where we hold relevant initial conditions fixed. Therefore, sunrises under MAL do not count as lucky, thereby avoiding the counterintuitive result provided by LCAL.

4. Modal Moral Luck

Pritchard (2006) also demonstrates how a modal understanding may illuminate cases of moral luck. More specifically, he shows how adopting MAL

---

[2] Pritchard (2015) has since argued that the significance condition should not play a role in luck. I disagree, so will continue using Pritchard's old account of luck.

[3] Levy (2011) argues that if we believe in agent causation (that an agent can create a new causal chain without being predetermined to do so) then there will be some cases of non-modally robust agent control.

for the purposes of moral luck more clearly pins down the intuition that one has been morally lucky. For example, let us consider our reckless truck driver case again, but let's further split the example such that in one case the reckless truck driver is driving on a usually quiet road and in another case the reckless truck driver is driving down a busy high street in the daytime. As Pritchard (2006, 6) states, although it is clear that the first truck driver has been unlucky to hit a pedestrian, it is a stretch to think that the second truck driver has been unlucky. The modal account captures the difference between the better-described examples. In the case of the quiet country road, there being a pedestrian in the road at the same time as our reckless truck driver may have occurred in the actual world but will not have occurred in a wide class of nearby possible worlds. In the case of the busy high street, our truck driver will have hit a pedestrian not only in the actual world, he will have hit the pedestrian in a wide class of nearest possible worlds as well. It's not obvious how an account of moral luck using LCAL can account for the difference in luck between these two cases if we think that the level control is equal in both the busy high street and on the country road. It seems, then, that by adopting MAL for moral luck allows us to have a more fine-grained understanding of why only certain cases where an agent is blameworthy or praiseworthy despite lack of control count as cases of moral luck. Stated in necessary and sufficient conditions, Pritchard's account of moral luck would look like the following:

(1) If an agent is morally lucky that event P, then p occurs in the actual world, and does not occur in a wide class of the nearest possible worlds where the relevant initial conditions for that event are the same as in the actual world.

(2) An event P is lucky for an agent S if P is significant for S.

However, Julia Driver (2013), in an attempt to create a modal account that better handles cases of moral luck, adds a contrastive element to Pritchard's account. That is, a luck attribution is about some P put in a relevant contrast to some other state of affairs. For example, for a simple lottery win case, I am lucky to win the lottery (p) in contrast to not winning the lottery (q). Driver provides the following formulation: "Event e is lucky or unlucky for a given individual in contrast to some other state of affairs (or, *rather than* some other state of affairs). An individual, S, is lucky that p rather than q" (2013, 9–10). Driver combines these two thoughts, the contrastive and the modal account of luck, to create the following account of luck:

(1) If an agent is lucky that event p rather than event q, then p occurs in the actual world, and does not occur in a wide class of the nearest possible worlds where the relevant initial conditions for that event are the same as in the actual world, whereas q does not occur in the actual world, and does occur in a wide

class of the nearest possible worlds where the relevant initial conditions for that event are the same as in the actual world.

(2) Whether or not p constitutes good luck or bad luck is relative to the interests of the agent (or the being with interests). (Driver 2013, 22)[4]

To show her account at work, and specifically how it differs from Pritchard's account, Driver provides the following example: "Sandra has had a narrow escape. She contracted an extremely rare, and extremely fatal, strain of flu. Fortunately, however, after two weeks of agonized suffering she has recovered and is recuperating in the hospital. Furthermore, through some odd and highly improbable combination of chemical factors the flu seems to have cured her arthritis. When her brother Bob comes to visit her she tells him happily: 'I am so lucky!' Bob disagrees with her, claiming that in reality she has been quite unlucky" (2013, 7–8).

In the example there is a contrast between Sandra catching the flu and being cured of her arthritis. That is, she is unlucky to have caught the flu, but at the same time lucky to have been cured of her arthritis, depending on the contrast that is made. This is the source of disagreement between Bob and Sandra; the former makes the contrast of Sandra not catching the flu, whereas the latter makes the contrast of Sandra being cured of arthritis. What Driver's account captures is the sense that an event can be both lucky and unlucky for the same agent, depending on what contrasts are being made.

Furthermore, the contrastivist addition serves to resolve any lack of clarity over what the relevant initial conditions are supposed to be. Driver uses an epistemic approach such that the relevant initial conditions are fixed by the reasonably foreseeable outcomes (foreseeable either by the agent or by the attributor). For example, in the Sandra case, Sandra is lucky to have caught the flu rather than suffer from arthritis, and she is unlucky to have caught the flu rather than not catching it all—in both instances what was reasonably foreseeable by Sandra is that she did not catch the flu, so the relevant set of possible worlds, Driver argues, are those where Sandra did not catch the flu. Sandra was lucky, as in the nearby set of relevant possible worlds (those worlds where what she can reasonably foresee) Sandra does not catch the flu at all.

For a case of moral luck, Driver asks us to consider the case of attempted murder where the would-be killer hits a bird rather than his intended victim. Driver states that what was reasonably foreseeable about this case is that the killer is successful. That is, the killer is lucky in this case, as in the relevant set of nearby possible worlds he is successful in the murder. Driver

---

[4] Driver takes an interests account of what is more generally named as the "significance" condition. This understanding is best described by Ballantyne (2011 and 2012).

adds that the contrast with what the agent finds foreseeable is also informative on the kind of reasons that the agent is responsive to, which in turn informs us about their character. She writes: "In the case of the attempted murder, the murderer is morally lucky because he hits a bird rather than his intended victim. He's done nothing wrong beyond the attempt. But again, this attempt is something that speaks badly of his character. In some nearby possible worlds (with the relevant conditions fixed, etc.) he has killed his intended victim. Again, the contrast—'rather than his intended victim'— demonstrates that he is not responding to the right sorts of reasons—he fails to value human life sufficiently. What is foreseeable, given the intentions, is the death of the intended victim" (2013, 23–24). So, Driver argues, the contrastivist addition to the modal account not only clears up what the relevant nearby possible worlds are but also, in cases of moral luck, what kind of character the person is.

With this modal account of moral luck in hand, we can now consider some problems with these accounts and use the solutions to these problems to create a different account of modal moral luck.

5. Problem 1: The Inclusivity Problem

The first problem with the current modal account of moral luck is that it is too inclusive. That is, it includes too many events as being morally lucky.

Consider a case where Emily saves Lauren from a burning building— an action that is easy enough for Emily, as she is extremely athletic. This is morally relevant, as Emily is morally praiseworthy for her actions. However, there had also been an earthquake that day near to where Emily makes the rescue. Had the earthquake been any closer (which it easily could have been), then Emily would have been unable to rescue Lauren from the fire due to the overwhelmingly unfriendly environment.

For now, let's assume that the contrast here is being made by a third-party attributor who makes the contrast that Emily was lucky to save Lauren, as in a relevant set of nearby possible worlds Emily failed to do so due to the earthquake being closer. That is, Emily was lucky to save Lauren rather than be affected by the earthquake—which would have changed the results of her action. If this is the contrast being made, then Emily qualifies as being morally lucky—she is morally praiseworthy for an event that could easily have gone otherwise. But that doesn't seem right at all. After all, Emily was in full control of saving Lauren despite the fact that the result of her action—the attempt at saving Lauren—could easily have been otherwise. The environmental fact that could easily have been otherwise does not seem as though it should have a bearing in this case on whether Emily has been morally lucky or not. She has been lucky in some sense, but not in the right sense that we should consider her *morally* lucky.

An initial response here might be to adjust the contrast such that it is Emily making the contrast rather than an informed third party. This would only work if Emily is ignorant of the facts about the earthquake. All that is required to generate the problem again is to stipulate that Emily is informed of the facts about the earthquake.

To press the problem further, consider another example, this time with moral blameworthiness in mind.

Due to an ongoing dispute, Sam plots to kill Craig. Lacking any imagination, Sam decides to simply walk to Craig's house and shoot him. Sam believes that Craig lives at No. 9 on his street, whereas Craig actually lives at No. 6. Sam walks down the street, gun in hand, and as he does so, heavy wind causes the top bolt in the "6" on Craig's door to fall out and the number to invert, giving the impression that Craig lives at No. 9. Sam sees the number 6 on the door, walks up, rings the bell, and shoots Craig.

In this case, Sam is morally blameworthy for his action of killing Craig. However, we can again make this a case of moral luck despite it intuitively not being so. The contrast here, and what a third party can reasonably foresee, is that Sam will walk to No. 9 and not kill Craig, rather than to No. 6 and kill Craig. So in a wide set of relevant nearby possible worlds, Sam does not kill Craig. So Sam is morally lucky that he killed Craig. Again, Sam has definitely been lucky in some way, but he doesn't seem morally lucky. Again, to explain the intuition, Sam was in full control of the fact that he shot Craig despite the fact that the result of his action—the murder attempt—could easily have been otherwise.

It is possible to make a formula for these kinds of cases. Take any event where an agent is in control of her actions, and the result of her actions will confer moral blameworthiness or praiseworthiness. Introduce some other fact in the causal vicinity that could easily have been otherwise and, had it been otherwise, would have prevented the action from happening at all, and now we have a case that although there is luck in the vicinity, we don't think it is a case of moral luck.

The overall problem here is that the current modal account doesn't pin down what it is to be specifically morally lucky.

6. An Action-Orientated Solution

To address this problem we can turn to some of the epistemological literature, specifically to how the work on epistemic luck has attempted to solve these kinds of problems. Pritchard (2005) has argued that certain kinds of luck are incompatible with knowledge and has provided an account of the kinds of luck, specifically epistemic luck, that affect knowledge. First, note that Pritchard's (2005) account of epistemic luck is *not* the following:

> S's true belief that P is lucky iff S believes that P in the actual world but not in relevant nearby possible worlds.

This formulation would run into the same kinds of problems as the modal moral account of luck in that too many cases of knowledge would be downgraded to simply lucky true belief. In other words, it would also be too inclusive. For example, imagine that you are walking down the street and luckily overhear two men talking about which horse will win in the horse race. We would not want to deny that you now know the information about the horse race (given that the two men are talking the truth), even though there is some luck in the territory. Instead, Pritchard (2005) states more precisely what it is that has to be lucky in order to prevent true belief from becoming lucky. The account is as follows:

> S's true belief that P is lucky iff S had formed the belief in the same way in relevant nearby possible worlds as she had in the actual world, then S's belief would no longer be true. (Pritchard 2005, 146)

In the horse race example, if S had formed the belief in the same way in nearby possible worlds—by overhearing the two men—then S's belief would still be true. By stating what precisely has to be lucky about beliefs in order for the luck to affect knowledge, Pritchard has avoided the problem of being overly inclusive. In effect, what Pritchard has done here is add a further world-fixing component. If we take into account Driver's contrastivism, not only do we need to hold fixed what is foreseeable in nearby possible worlds, we also need to hold fixed the belief formation across these worlds so that we can precisely assess whether the agent has been subject to the kind of luck that would be knowledge undermining. In the same manner, we can state what precisely has to be lucky in cases of moral luck in order to qualify as a case of moral luck by adding a further world-fixing component. My suggestion here will be that this further fixing component should be the action performance, as this seems to give the best results. So the account would look as follows:

S is morally lucky that E iff

(1) S's action had been performed in the same way as in the actual world but the results (E) would have been different in a wide set of relevant nearby possible worlds.
(2) Whether or not E constitutes good luck or bad luck is relative to the interests of the agent (or the being with interests).

This account solves the problematic cases used above. In the case of Emily saving Lauren, given that we hold Emily's performance of the rescue fixed, it was not lucky that Emily rescued Lauren. If we take into account the nearby earthquake, then we are no longer holding Emily's performance fixed, as this would have changed the way Emily performed the action. By fixing the performance of the action, we also fix certain other facts around the case regardless of their own modal robustness. So in any nearby possible worlds where we hold fixed the performance of Emily's action, the earthquake also does not occur, as this would have affected the performance in such a way that the same results would not have occurred. In one sense, Emily has been lucky to save Lauren, but so long as we think moral luck depends on the fixing of action performance, Emily saving Lauren is not a case of moral luck.

In the same manner, in the case of Sam and Craig, given that Sam performed the action of going up to Craig's house and shooting him, he was not lucky. If we hold the facts of Sam's performance fixed, then this fixes the fact that Sam went to Craig's house regardless of other kinds of luck lurking in the causal history. Despite the Gettier-like nature of the example, given that holding Sam's performance fixed requires that he reaches Craig's front door, the only relevant nearby possible worlds that we can use for assessing whether Sam has been morally lucky are those worlds where he reaches Craig's door. In a wide set of those worlds, Sam succeeds in shooting Craig. So Sam is not morally lucky, despite the fact that there is some epistemic luck in the example.

What about uncontroversial cases of moral luck, such as the reckless truck driver examples? In those cases we hold fixed the action performance of driving recklessly, but in one case this action results in the death of a person and in the other it does not. The result of hitting the pedestrian is a result that would not have occurred in a wide set of relevant nearby possible worlds, or in contrastive terms it was unlucky that, given the truck driver was driving recklessly, the truck driver hit the pedestrian rather than, given the truck driver was driving recklessly, the truck driver did not hit a pedestrian. In this way, the action-orientated account preserves clear cases of moral luck, but it excludes cases where moral action has occurred and there is luck in the vicinity but where this luck is irrelevant for being morally lucky.

There is, however, still another problem, both for this account and for Driver's original account.

7. Problem 2: The Significance Problem

The second problem for both Driver's account and the action-orientated account of luck concerns the second condition for luck: the significance condition.

The significance condition for luck is there to capture the thought that lucky events must go well or badly for the agent in question (Coffman 2006). A chancy event that affects no one or does not affect someone in a positive or negative way is not thought to be a lucky event. We might, however, think that one of the further functions of the significance condition of luck is to pick out whether the event in question has been a case of good luck or bad luck. That is, not only should we have a significance condition to capture the thought that a lucky event has a value attached to it for the agent involved, the significance condition should also be informative about what value that event has. So far, the significance condition has been understood in terms of the interests of the agent. Ballantyne provides the following understanding for what this kind of significance condition should look like: "Individual X is lucky with respect to E only if (i) X has an interest N and (ii) E has some objectively positive or negative effect on N (in the sense that E is good for or bad for X)" (2012, 13). For example, Amy is lucky to win the lottery, as she has an interest in gaining lots of money and winning the lottery has a positive effect on this interest (in the sense that winning the lottery is good for satisfying this interest).

With these thoughts in mind, this leads us to a problem with the following example: There are two reckless truck drivers, driving down equally quiet roads. The chances of either driver hitting a pedestrian are small. The first truck driver hits no one. The second truck driver hits a pedestrian. However, the second truck driver has a sadistic nature. Despite not intentionally hitting the pedestrian, the truck driver gets lots of pleasure out of the incident.[5]

In this case, the truck driver has committed a morally bad action where the results are modally non-robust, so the prediction of the account should be that the truck driver is morally unlucky. However, given that the significance condition measures the value of the luck (good or bad luck) and that it does this only by looking at the interests of the agent, the account will hold that the truck driver has been subject to moral good luck. That is, a morally relevant action has occurred where the results are modally non-robust but also where the results are in the interests of the agent. More precisely, given that the sadistic truck driver has an interest in hurting others and that hitting a pedestrian has been good for satisfying this interest, and that the modal conditions have been met, the truck driver has been subject to good luck. This doesn't seem to be correct when assessing how the agent has been morally lucky. We should think the agent has been morally unlucky regardless of whether the results give the agent pleasure or not.

The problem also works the other way round. To do that, we need to change the case to where the results of an action will be morally bad and likely to occur but the results of the action fail to be morally bad despite the intentions of the agent. An example of this would be an attempted murder where the killing fails due to some modally non-robust circumstances. In these cases, the agent's interests have been frustrated. According to the current understanding of the significance condition, the agent has suffered a case of bad luck. Morally speaking, however, the agent has been affected by moral good luck.

---

[5] Note that intentions might not matter. Just because someone does something intentionally does not mean that he was non-lucky in achieving his ends. One can use radically unreliable methods yet still may achieve one's ends. In the case of the sadistic truck driver, however, it looks as though we should probably think of the killing of a pedestrian as being unintentional, despite being pleasurable to the truck driver.

The problem, then, is that under our current understanding of the significance condition, the current account of moral luck can potentially pick out the wrong value for the moral luck that has occurred.[6]

8. Solution: Relativising the Significance Condition

The problem here is that the significance condition is picking out not the moral value of the action but the value of the outcome of the action orientated around the agent's interests. So the remedy to the problem is to adjust what the significance condition picks out. To make sure that it is only the relevant domain that we are picking out with the significance condition, we need to relativise the significance condition to the normative domain to which we are making the luck attribution. The normative domain in this case is the moral domain, so the significance here needs to be moral significance rather than interests significance.

The way that the results of an action are morally significant may depend on one's moral theory. To use our sadistic truck driver example, a utilitarian may hold that the action of hitting the pedestrian is morally significant for the truck driver because the truck driver has decreased overall utility. A virtue theorist, on the other hand, may hold that the action is morally significant, as it reinforces the vicious traits of the truck driver.

Significance might be a confusing term to use here, particularly when referring to moral significance. A better term might be value. The original interests condition was there to precisify the kind of value that an event has to have in order to be lucky for an agent. Here we are doing the same work, but, rather than for luck in general, for luck in a specific normative domain. Again, the moral value of an event will depend on the kind of moral theory one subscribes to. What moral theory we apply here does not matter, so long as it has an element of externalism that allows for results of actions to have some moral value. If this is right then our account of moral luck should look as follows:

S is morally lucky that E iff

(1) S's action had been performed in the same way as in the actual world but the results (E) would have been different in a wide set of relevant nearby possible worlds.

(2) The results (E) of S's action are of positive or negative moral value (where the moral value is defined by the moral theory that we are using).

With this account in hand, we can say something more accurate of the malicious truck driver. He is the victim (in the loosest possible sense) of

---

[6] Thanks to Alan Wilson and Steven Hales for suggesting this problem and assisting me in getting to grips with it.

moral bad luck. His actions satisfy condition (1) by virtue of the modal non-robustness of hitting a pedestrian and his actions satisfy condition (2) because the results of these actions were morally valuable (negative), at least according to a utilitarian theory of ethics (more pain has been caused than pleasure). We can state that the value of the luck in this case was bad, as the ethical value of the action was negative.

One problem that might arise with this different understanding of significance concerns the fact that a lucky event is lucky *for* an agent. However, the account above does not specify that the results of the action have been lucky for that agent, only that some luck has occurred and that an agent was involved. This is especially the case if we have a utilitarian or consequentialist understanding of moral value. The agents in question may have decreased or increased overall utility, but they haven't increased or decreased overall utility for themselves.

To answer this objection requires an understanding of role reversal of the two conditions for luck. In the general modal account of luck, it is the significance condition that ties the event agent in such a way that the event can be said to be lucky for that agent. That is, modally non-robust event E occurs, and E affects S in a positive or negative way.

However, what ties the agent to the event in the account of modal moral luck above is not the value of the event but the fact that the action was carried out by the agent. The action performance is what ties the agent to the event in such a way that we can state that S has been morally lucky (rather than that moral luck has just occurred). What makes an S morally lucky is that she has carried out an action that fulfils the conditions set above, not that the value of the results of the action has affected her somehow (the results may in no way affect the agent).

If all this is agreeable, then the above account, with contrastivist considerations of Driver, should be considered the correct account of resultant moral luck.

This essay, then, has demonstrated two problems with the current modal account of resultant moral luck: the problem of inclusivity and the significance problem. I have provided solutions to both of these problems, resulting in the action-orientated account of modal resultant moral luck. This account of moral luck should be considered the correct account of moral luck.

References

Coffman, E. J. 2006. "Thinking About Luck." *Synthese* 158, no. 3:385–98.
Ballantyne, Nathan. 2011. "Anti-Luck Epistemology, Pragmatic Encroachment, and True Belief." *Canadian Journal of Philosophy* 41, no. 4:485–503.
———. 2012. "Luck and Interests." *Synthese* 185, no. 3:319–34.

Driver, Julia. 2013. "Luck and Fortune in Moral Evaluation." In *Contrastivism in Philosophy*, ed. Martijn Blaauw, 154–73. London: Routledge.

Hanna, Nathan. Forthcoming "Moral Luck Defended." *Nous*.

Lackey, Jennifer. 2008. "What Luck Is Not." *Australasian Journal of Philosophy* 86:255–67.

Latus, Andrew. 2000. "Moral and Epistemic Luck." *Journal of Philosophical Research* 25:149–72.

Levy, Neil. 2011. *Hard Luck: How Luck Undermines Free Will and Moral Responsibility*. New York: Oxford University Press.

Nagel, Thomas. 1993. "Moral Luck." In *Moral Luck*, ed. Daniel Statman, 57–71. Albany: State University of New York Press.

Pritchard, Duncan. 2005. *Epistemic Luck*. Oxford: Oxford University Press.

———. 2006. "Moral and Epistemic Luck." *Metaphilosophy* 37, no. 1:1–25.

———. 2015. "The Modal Account of Luck." Included in this collection.

Statman, Daniel. 1991. "Moral and Epistemic Luck." *Ratio* 4:146–56.

Williams, Bernard. 1981. *Moral Luck*. New York: Cambridge University Press.

———. 1993. "Postscript." In *Moral Luck*, ed. Daniel Statman, 241–47. Albany: State University of New York Press.

Zimmerman, Michael. 1987. "Luck and Moral Responsibility." *Ethics* 97:374–86.

INDEX

The Philosophy of Luck, First Edition. Edited by Duncan Pritchard and Lee John Whittington.
Chapters © 2014 Metaphilosophy LLC and John Wiley & Sons Ltd, except for "Luck as Risk
and the Lack of Control Account of Luck" © 2015 Metaphilosophy LLC and John Wiley &
Sons Ltd. Book compilation © 2015 Metaphilosophy LLC and John Wiley & Sons Ltd.
Published 2015 by John Wiley & Sons Ltd.